Various quadrilaterals
Koginzashi Embroidery
Hirosaki City, Aomori Prefecture
Tokyo Budo-kan
Adachi, Tokyo.
The great Hanshin-Awaji Earthquake Memorial Disaster Reduction and Human Renovation Institution
Kobe City, Hyogo Prefecture
1
2
3
4
Vaulting Horse
Hishimochi Sweet
Cable car
Kawanishi City,
Hyogo Prefecture

Table of contents

Hiroto

Nanami

Let's learn mathematics together.

Daiki

Yui

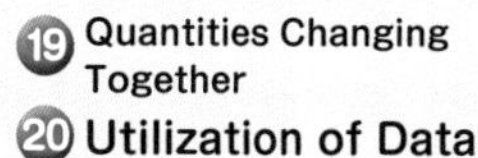

3 learning competencies to develop.

1 Thinking Competency

Competency to think the same or similar way

Competency to find what you have learned before and think in the same or similar way

Competency to find rules

Competency to analyze numbers and various expressions that change together, and investigate any rules

Competency to explain the reason

Competency to explain the reason why, based on learned rules and important ideas

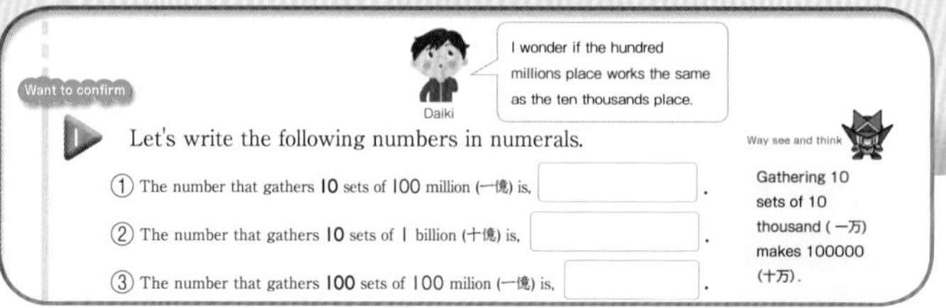

Want to confirm

Daiki: I wonder if the hundred millions place works the same as the ten thousands place.

Let's write the following numbers in numerals.

① The number that gathers 10 sets of 100 million (一億) is, ☐ .

② The number that gathers 10 sets of 1 billion (十億) is, ☐ .

③ The number that gathers 100 sets of 100 milion (一億) is, ☐ .

Way see and think: Gathering 10 sets of 10 thousand (一万) makes 100000 (十万).

2 Judgement Competency

Competency to find mistakes

Competency to find the over generalization of conventions and rules

Competency to categorize by properties and patterns

Competency to categorize numbers and shapes by observing their properties and patterns

Competency to compare ideas and ways of thinking

Competency to find the same or different ways of thinking by comparing friends' ideas and your own ideas

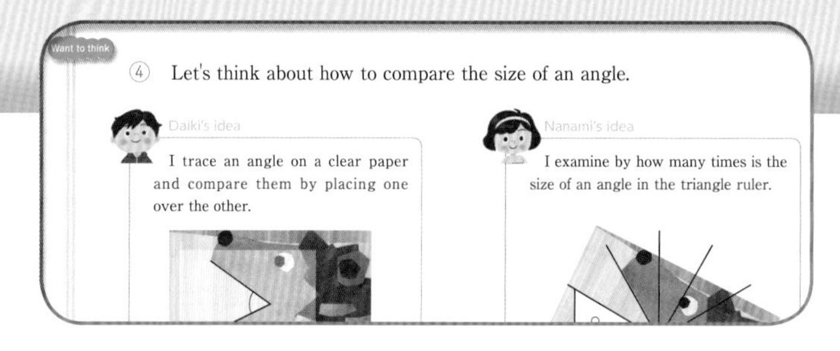

Want to think

④ Let's think about how to compare the size of an angle.

Daiki's idea: I trace an angle on a clear paper and compare them by placing one over the other.

Nanami's idea: I examine by how many times is the size of an angle in the triangle ruler.

3 Representation Competency

Competency to represent sentences with diagrams or expressions

Competency to read problem sentences, draw diagrams, and represent them using expressions

Competency to represent data in graphs or tables

Competency to express the explored data in tables or graphs in simpler and easier ways

Competency to communicate with friends and yourself

Competency to communicate your ideas to friends in simpler and easier manners and to write notes in simpler and easier ways for you

The figure on the right shows the process Daiki followed to find the number of seconds passed since Yui's birth. Let's explain Daiki's idea.

Daiki's notebook

$60 \times 60 = 3600$

$3600 \times 24 = 86400$

$86400 \times 365 = 31536000$

$31536000 \times 10 = 315360000$

Want to deepen

Let's find monsters.

Monsters which represent ways of thinking in mathematics

Setting the unit.

Once you have decided one unit, you can represent how many.

If you try to separate...

Decomposing numbers by place value and dividing figures make it easier to think about problems.

If you represent in other way...

If you represent in other expression, diagram, table, etc., it is easier to understand.

Can you do the same or similar way?

If you find something the same or similar, then you can understand.

You wonder why?

Why this happens? If you communicate the reasons in order, it will be easier to understand for others.

If you try to arrange...

You can compare if you align the number place and align the unit.

If you try to summarize...

It is easier to understand if you put the numbers in groups of 10 or summarize in a table or graph.

If you try to change the number or figure...

If you try to change the problem a little, you can better understand the problem or find a new problem.

Is there a rule?

If you examine a few examples, then you can find out whether there is a rule.

Ways to think learned in the 3rd grade.

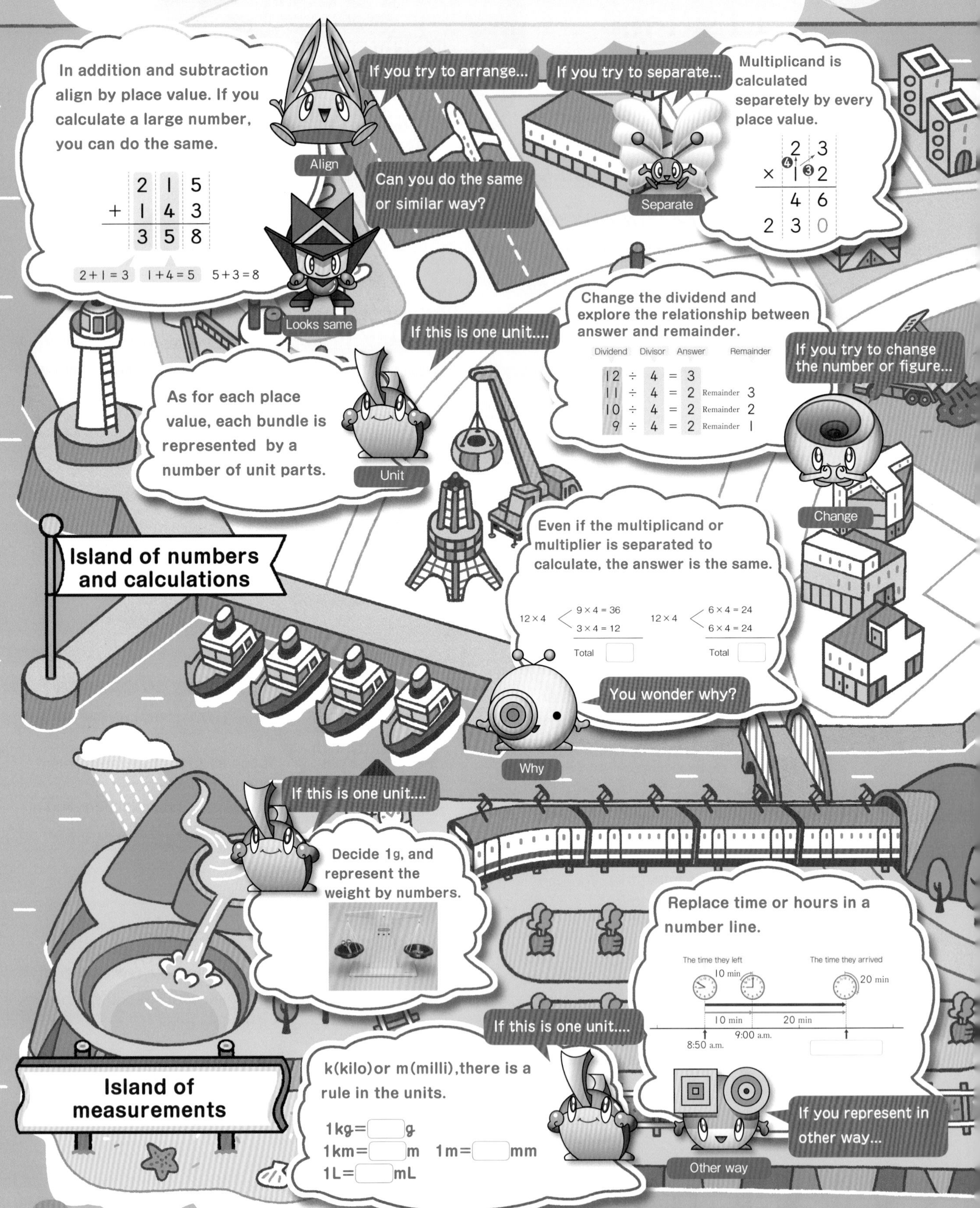

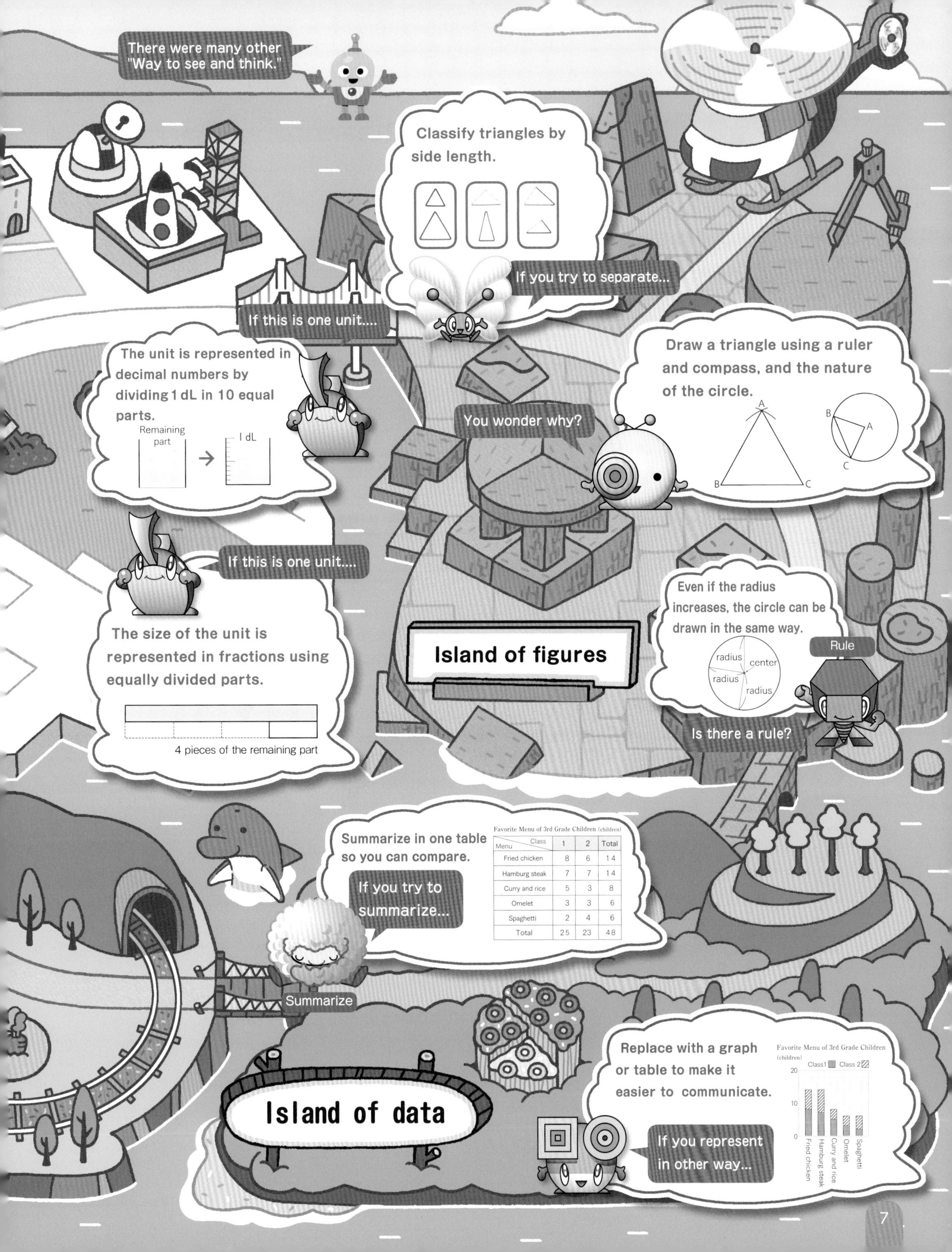

Favorite Menu of 3rd Grade Children (children)

Menu \ Class	1	2	Total
Fried chicken	8	6	14
Hamburg steak	7	7	14
Curry and rice	5	3	8
Omelet	3	3	6
Spaghetti	2	4	6
Total	25	23	48

Let's progress through 3 ways of learning.

1 Self directed learning: Learning on your own initiative.

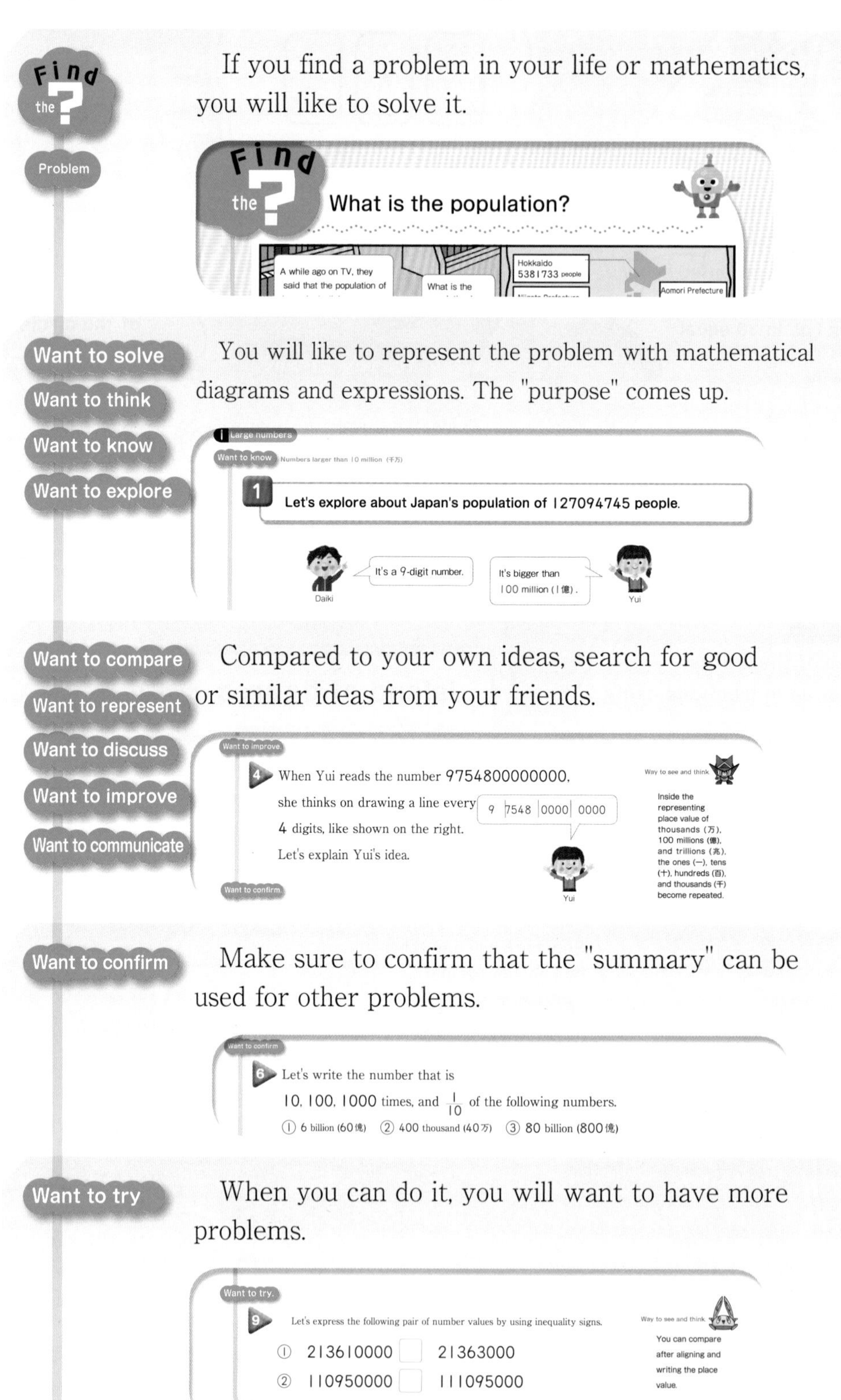

If you find a problem in your life or mathematics, you will like to solve it.

You will like to represent the problem with mathematical diagrams and expressions. The "purpose" comes up.

Compared to your own ideas, search for good or similar ideas from your friends.

Make sure to confirm that the "summary" can be used for other problems.

When you can do it, you will want to have more problems.

2 Dialogue learning: Learning together with friends.

As learning progresses, you will want to know what your friends are thinking. Also, you will like to share your own ideas with your friends. Let's discuss with the person next to you, in groups, or with the whole class.

3 Deep learning: Usefulness and efficiency of what you learned.

Let's cherish the feeling "I want to know more." and "Can I do this in another case?" Let's deepen learning in life and mathematics.

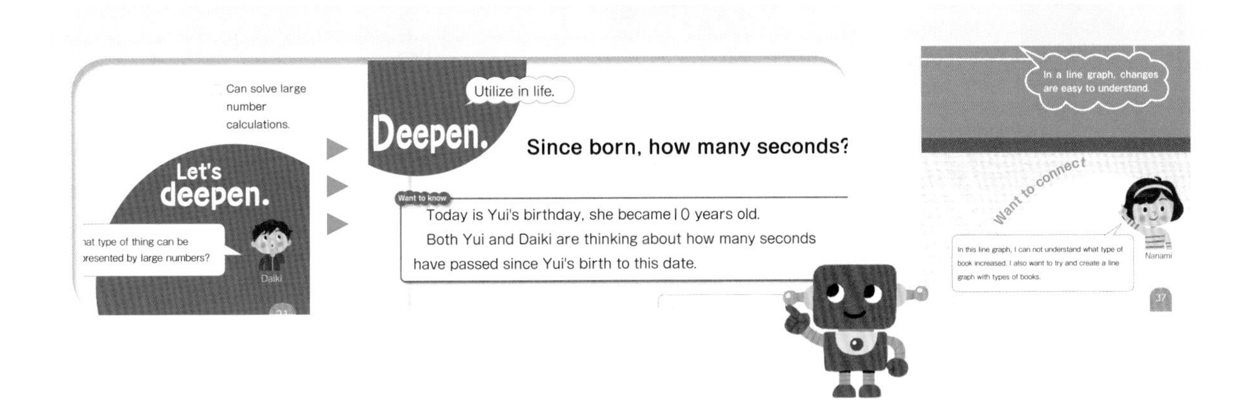

What is the population?

Problem As the population of Japan is 127094745 people, I wonder how many people that should be.

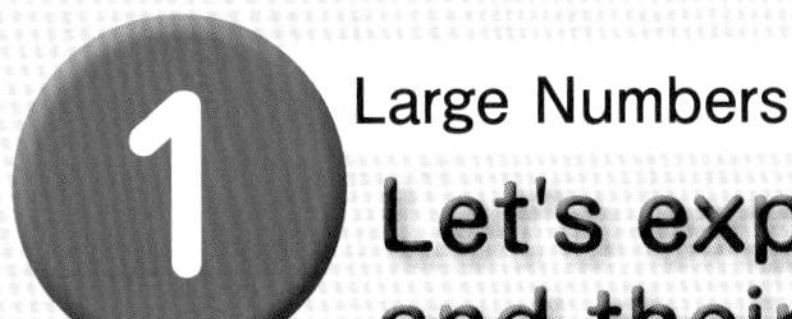

Large Numbers

Let's explore how to express numbers and their structure.

1 Large numbers

Want to know Numbers larger than 10 million (千万)

Let's explore about Japan's population of 127094745 people.

Purpose As for Japan's population, how is the correct way to read it?

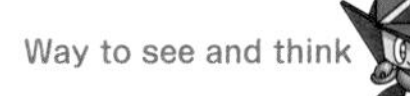

If a number gathers 10 sets, it becomes one place higher.

① As for the place with a 2, what place value is it?

② How many sets of 10 million (千万) you need to gather, in order to reach the next higher place?

The number that gathers 10 sets of 10 million (千万) is written as 100000000, and read as one **hundred million** (一億). Also written as 100 million (1億). The place value that follows the ten millions place is the **hundred millions place**.

Want to represent

③ Let's read the population of Japan.

Millions			Thousands			Ones			
100 millions	10 millions	Millions	Hundred thousands	Ten thousands	Thousands	Hundreds	Tens	Ones	
一億	千	百	十	一万	千	百	十	一	
1	2	7	0	9	4	7	4	5	pepole

100 million (一億) and 10 thousand (一万) represent one unit.

Summary

127094745 is read "one hundred twenty-seven million ninety-four thousand seven hundred forty-five." Also it is written as 127 094 745 (1億2709万4745).

I wonder if the hundred millions place works the same as the ten thousands place.

Want to confirm

1 Let's write the following numbers in numerals.

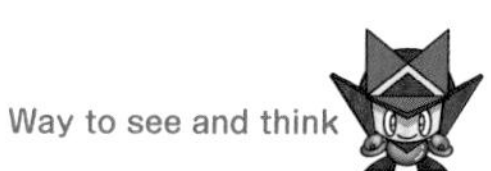

Gathering 10 sets of 10 thousand (一万) makes 100000 (十万).

① The number that gathers 10 sets of 100 million (一億) is, [ ].

② The number that gathers 10 sets of 1 billion (十億) is, [ ].

③ The number that gathers 100 sets of 100 milion (一億) is, [ ].

Want to extend

2 Let's write and read the population of U.S.A, China, Brazil, and the world.

It's easy to start writing from the ones place in the table.

	Billions			Millions			Thousands			Ones				
	100 billions	10 billions	Billions	100 millions	10 millions	Millions	Hundred thousands	Ten thousands	Thousands	Hundreds	Tens	Ones	
	千	百	十	一 億	千	百	十	一 万	千	百	十	一	
U.S.A.				3	2	4	4	5	9	0	0	0	people
China													people
Brazil													people
World													people

7550000000 pepole World's population

46354000 pepole Spain

1409517000 pepole China

324459000 pepole U.S.A.

127095000 pepole Japan

209288000 pepole Brazil

49700000 pepole Kenya

24451000 pepole Australia

Want to find in our life

3 From your surroundings, let's look for a number that uses the hundred millions place.

Want to know Numbers larger than 100 billion

Yokohama City, Kanagawa Prefecture

2

The following amount of money is the import value of Yokohama Port in 2016, 3 7999 0000 0000 yen.

Let's explore this number.

Daiki

Yui

Purpose How should we read 3 7999 0000 0000?

① As for the place with a 7, what place value is it?

② How many sets of 100 billion (千億) do you need to gather, in order to reach the next higher place?

The number that gathers 10 sets of 100 billion (千億) is written as 1000000000000 and read as **one trillion** (一兆). Also written as 1 trillion (1兆).

③ Let's read the import value of Yokohama Port.

	Billions			Millions			Thousands			Ones		
Trillions	100 billions	10 billions	Billions	100 millions	10 millions	Millions	Hundred thousands	Ten thousands	Thousands	Hundreds	Tens	Ones
一	千	百	十	一	千	百	十	一	千	百	十	一
兆				億				万				
3	7	9	9	9	0	0	0	0	0	0	0	0

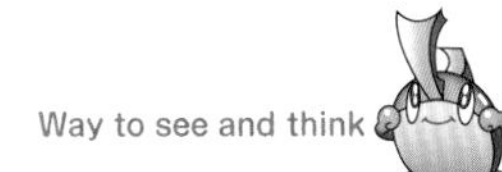

1 trillion (一兆) and 100 million (一億) represent one unit.

Summary

3 7999 0000 0000 is read "three trillion seven hundred ninety-nine billion and nine hundred million." Also it is written as 3 799 900 000 000 (3兆7999億).

④ Let's try to expand the table on exercise ③ until it reaches the quadrillions place.

Want to improve

When Yui reads the number 9754800000000, she thinks on drawing a line every 4 digits, like shown on the right.

Let's explain Yui's idea.

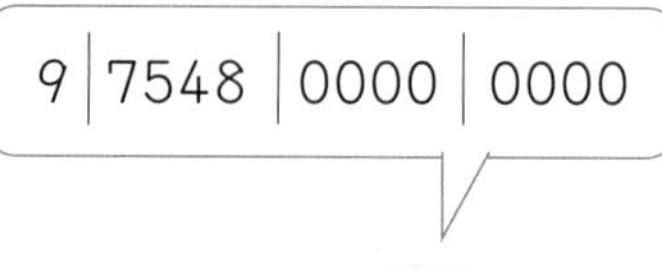

Way to see and think

Inside the representing place value of thousands (万), 100 millions (億), and trillions (兆), the ones (一), tens (十), hundreds (百), and thousands (千) become repeated.

Want to confirm

Let's read the following numbers.

① 9 4600 0000 0000 km

(Distance which light travels in one year)

② 4078 0000 0000 0000 km

(Distance between the Earth and the North Star)

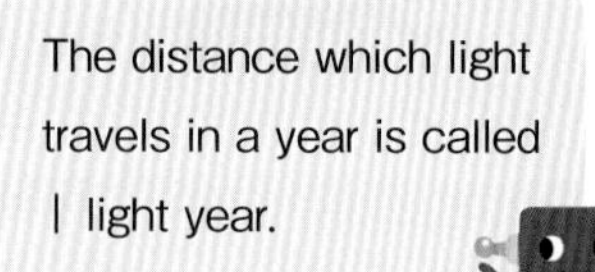

That's it

Number grouped by 3 digits

As for large numbers in Japan, every 4 digit place value has its own way of reading.

9387	4160	2571	0364
兆	億	万	

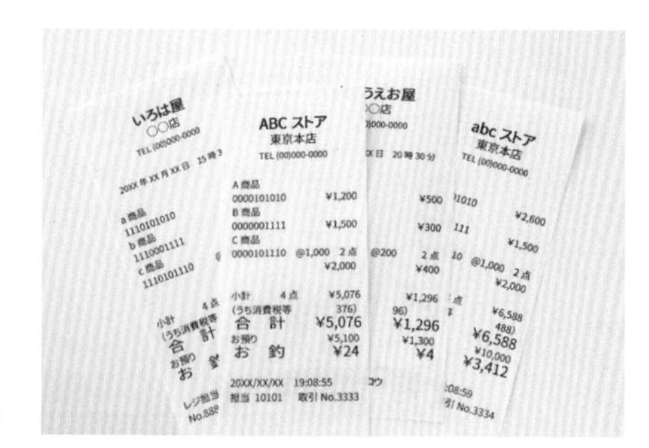

But, in our surroundings, there are many things which are divided by a "," every 3 digits.

In other countries such as U.S.A., every 3 digit place value has its own way of reading.

9, 387, 416, 025, 710, 364

Want to represent

Way to see and think 

As for a number that gathers 10 sets, is the same as multiplying by 10. As for $\frac{1}{10}$ of a number is the same as dividing by 10.

3 **Let's write and read the numbers that are 10, 100, and 1000 times of 325 6900. Also, let's write and read the number which is $\frac{1}{10}$ of 325 6900.**

Purpose

How does the place value of a number change when getting 10 times or $\frac{1}{10}$ of the number?

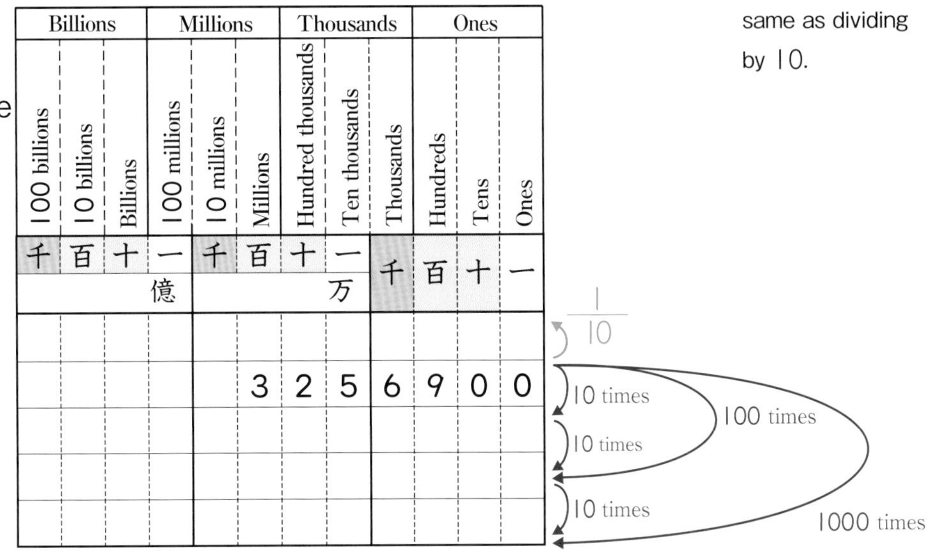

Billions				Millions			Thousands			Ones	
100 billions	10 billions	Billions	100 millions	10 millions	Millions	Hundred thousands	Ten thousands	Thousands	Hundreds	Tens	Ones
千	百	十	一	千	百	十	一	千	百	十	一
			億				万				
					3	2	5	6	9	0	0

Summary

In any whole number, when getting 10 times, the place value increases one position. Also, when getting $\frac{1}{10}$, the place value decreases one position.

Want to confirm

6 Let's write the number that is 10, 100, 1000 times, and $\frac{1}{10}$ of the following numbers.

① 6 billion (60億) ② 400 thousand (40万) ③ 80 billion (800億)

7 Let's write the following numbers in numerals.

10 times 100 million (1億), ☐.

100 times 100 million (1億), ☐.

1000 times 100 million (1億), ☐.

10000 times 100 million (1億), ☐.

Want to represent

8 In the following number lines, let's fill in the ☐ with numbers.

① 0 ☐ 50 million ☐ 100 million ☐

② 0 1 billion ☐ ☐ 10 billion ☐

③ 0 ☐ ☐ 700 billion 1 trillion ☐

Way to see and think

How much does one scale represent?

Want to try

9 Let's fill in the ☐ with inequality signs.

① 213610000 ☐ 21363000

② 110950000 ☐ 111095000

Way to see and think

You can compare after aligning and writing the place value.

Develope

That's it

Numbers with place value higher than 1 quadrillion (1000兆)

1 quadrillion (1000兆) is a very large number. For example, if you count one number each second, it will take you about 30 million (3000万) years to count from 1 to 1 quadrillion (1000兆).

Additionally, there are place values higher than 1 quadrillions (1000兆) place. If you write the number character for "Mu-ryo-tai-su", 68 zeros are aligned after 1.

100000000000000000000000000000000000000000000000000000000000000000000

mu-ryou-tai-suu
Fukashigi
nayuta (novemdecillion)
a-sou-gi
gou-ga-sya: from Gou-ga-sya the number system changes 8-digit system instead of 4-digit system
1 quindecillion
100 tredecillion (no quattuordecillion)
10 duodecillion
1 undecillion
100 nonillion (no decillion)
10 octillion
1 septillion
100 quintillion (no sextillion)
10 quadrillion
1 trillion
100 million (no billion)
10 thousand (no 100 thousand)
Thousand
Hundred
Tens
Ones

2 Structure of whole numbers

Want to solve

1 **Using the ten cards from** 0 **to** 9**, let's create 10-digit whole numbers.**

	Millions			Thousands			Ones		
Billions	100 millions	10 millions	Millions	Hundred thousands	Ten thousands	Thousands	Hundreds	Tens	Ones
十	一	千	百	十	一	千	百	十	一
	億				万				

Daiki: Can not use 0 in the billions (十億) place.

Nanami: On each place value, one number character can be inserted.

You can make a lot of whole numbers.

Purpose Which characteristic do the whole numbers have?

① Let's make the largest number.

② Let's make the smallest number.

Summary

Any whole number, no matter how big it is, can be written and represented by using the following 10 numerals: 0, 1, 2, 3, 4, 5, 6, 7, 8, 9.

Want to represent

1 Concerning 30 9800 0000 0000, let's fill in the ☐ with numbers.

① It is the sum of 30 sets of 1 trillion (1兆) and ☐ sets of 100 million (1億).

② It is the sum of ☐ sets of 10 trillion (10兆), ☐ sets of 100 billion (1000億), and 8 sets of 10 billion (100億).

③ It is a number that gathers ☐ sets of 100 million (1億).

Way to see and think

As in 1 trillion (1兆) and 100 million(1億), the representation of whole numbers change depending on what is taken as a unit.

Trillions				Billions			Millions			Thousands			Ones		
Quadrillions	100 trillions	10 trillions	Trillions	100 billions	10 billions	Billions	100 millions	10 millions	Millions	Hundred thousands	Ten thousands	Thousands	Hundreds	Tens	Ones
千	百	十	一	千	百	十	一	千	百	十	一	千	百	十	一
			兆				億				万				
		3	0	9	8	0	0	0	0	0	0	0	0	0	0
10 millions	Millions	Hundred thousands	Ten thousands	Thousands	Hundreds	Tens	Ones								

3 Calculation of large numbers

Want to think

1 **The table on the right shows the number of copies of magazines that have been published in Japan this year. Let's think about this numbers.**

Number of published magazines in Japan

Type	Number of copies
Weekly magazine	700000000
Monthly magazine	1800000000

① Find the sum of the number of copies of magazines published weekly and monthly. Let's compare the following two math expressions.

ⓐ 700000000 + 1800000000

ⓑ 700 million + 1 billion 800 million

Way to see and think

As for 700000000, if 100000000 is one unit, then it has 7 units. It can be represented as 700 million (7億).

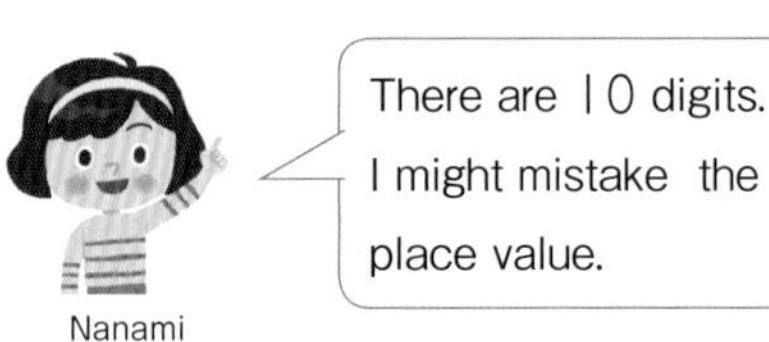

② How many copies is the difference between the number of monthly and weekly publication magazines?

The result of adding numbers is called **sum**. The result of subtracting numbers is called **difference**.

Want to confirm

1 Let's find the following sums and differences.

① Sum of 1 billion 700 million and 2 billion 900 million.

② 2 million 350 thousand + 5 million 150 thousand

③ Difference of 23 trillion and 8 trillion.

④ 80 billion 700 million − 69 billion 200 million

Words

【和】(Sum) To combine together.

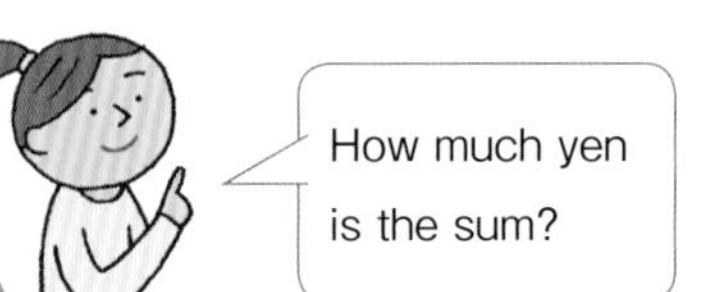

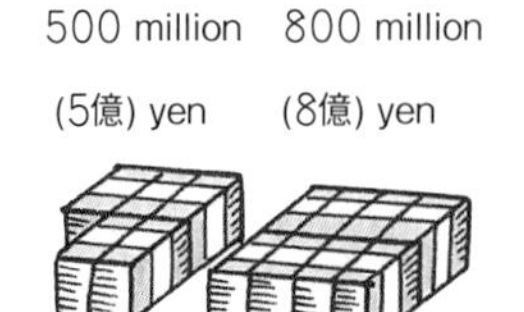

【差】(Difference) Take away.

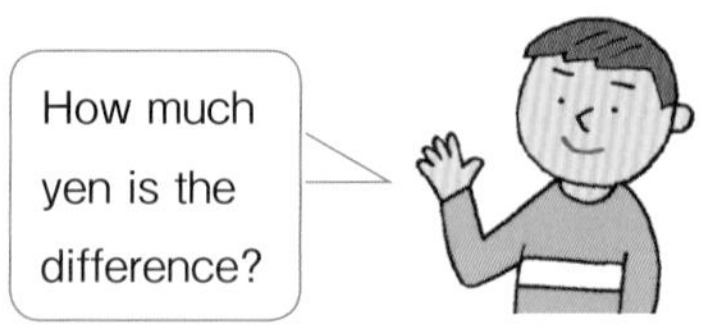

Want to think

2 Let's think about the following problem.

① Yui's city library has a monthly budget of 650000 yen to purchase books. How much is the annual budget?

It is easier to understand if 650000 is represented as 650 thousand (65万).

Yui

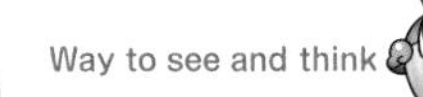

Way to see and think

As for 650000, if 10000 is one unit, then it can be represented as 650 thousand (65万).

Math Sentence:

② Rin's school budget to buy lunch for 5 days is 350000 yen. How much is the daily budget?

Math Sentence:

The result of multiplying numbers is called **product**. The result of dividing numbers is called **quotient**.

Want to confirm

3 Let's find the following products and quotients.

① 760 thousand $\times 2$ ② 26 billion 400 million $\times 10$

③ 8 million 500 thousand $\div 10$ ④ 90 trillion $\div 9$

Product
【積】 Pile up, multiply.

Quotient
【商】 Measure. Compare.

What you can do now

Can read large numbers.

1 Let's read the following numbers.

① The distance from the Sun to the Earth. 149600000 km

② Total budget of Japanese Government (in 2017). 97450000000000 yen

Understanding how large numbers work.

2 Let's fill in the ☐ with a number or a word.

① The number that gathers 10 sets of 10 million is ☐.

Also, the number that gathers 10 sets of 100 billion is ☐.

② 100 million is ☐ sets of 10 thousand. Also, 1 trillion is ☐ sets of 100 million.

③ The digit 7 in the number 72000000000000 represents 7 sets of ☐.

④ The number that is 10 times 180 billion is ☐.

⑤ The number that is $\frac{1}{10}$ of 23 trillion is ☐.

Understanding how to represent numbers.

3 Let's write the following numbers in numerals and read them.

① The sum of 46 sets of 1 trillion and 2375 sets of 100 million.

② The sum of 20 sets of 10 trillion and 45 sets of 10 billion.

Can compare the size of numbers.

4 Let's fill in the ☐ with inequality signs.

① 9860124700 ☐ 98600124700

② 7049082513 ☐ 7049802513

Can solve large number calculations.

5 Let's solve the following calculations.

① 38 billion 300 million + 42 billion 900 million

② 73 million 510 thousand − 3 million 960 thousand

③ 5 million 260 thousand × 5

④ 7 billion 200 million ÷ 8

Supplementary problems ▶ p.149

Usefulness and efficiency of learning

1 Let's read the following numbers.

① The Sun's diameter: 1390000000m

② Amount of rice produced (in 2017): 7306000000kg

☐ Can read large numbers.

2 Let's fill in the ☐ with a number or a word.

① The number 6 in 36495000000 is in the ☐ value place.

② As for the number 465 billion, gathers ☐ sets of 1 billion.

③ The number 1 trillion is equal to ☐ times 10 billion.

☐ Understanding how large numbers work.

3 Let's write the following numbers in numerals.

① The number that is 100 times 340 million.

② The sum of 3 sets of 1 trillion and 48 sets of 100 million.

③ The number that gathers 58013 sets of 100 million.

☐ Understanding how to represent numbers.

4 Let's make various numbers by using all the 14 cards shown on the right.

① Let's make the largest number.

② Let's make the smallest number.

0	0	0	0	0
1	2	3	4	5
6	7	8	9	

☐ Can compare the size of numbers.

5 Let's solve the following calculations.

① $2400000000 + 4500000000$

② $38000000000000 - 16000000000000$

③ 750000×12

④ $640000 \div 8$

☐ Can solve large number calculations.

Let's deepen.

What type of thing can be represented by large numbers?

Daiki

Since born, how many seconds?

Want to know

Today is Yui's birthday, she became 10 years old.

Both Yui and Daiki are thinking about how many seconds have passed since Yui's birth to this date.

Today I became exactly 10 years old, how many seconds are there since my birth?

It's better to think in order. First, 1 minute = 60 seconds. As 1 hour = 60 minutes, $60 \times 60 = 3600$, therefore, 1 hour has 3600 seconds.

Want to express

The figure on the right shows the process Daiki followed to find the number of seconds passed since Yui's birth. Let's explain Daiki's idea.

Daiki's notebook

$60 \times 60 = 3600$

$3600 \times 24 = 86400$

$86400 \times 365 = 31536000$

$31536000 \times 10 = 315360000$

Want to deepen

Between which ages, there will be more than 1 billion (10 億) seconds passed since Yui was born? Let's choose from Ⓐ~Ⓓ and use expressions and words to explain your reasons.

Ⓐ Between 10 and 20 years old. Ⓑ Between 20 and 30 years old.

Ⓒ Between 30 and 40 years old. Ⓓ Between 40 and 50 years old.

The representation of seconds since the time I was born became a very large number.

Comparing temperatures?

Niigata City's average temperature per month

month	1	2	3	4	5	6	7	8	9	10	11	12
temperature (℃)	3	3	6	12	17	21	25	27	23	16	11	6

Auckland City's average temperature per month

month	1	2	3	4	5	6	7	8	9	10	11	12
temperature (℃)	20	20	18	16	14	12	11	12	13	14	16	18

Problem Except for bar graphs, what other graphs are there?

2 Line Graph

Let's explore graphs to represent changes.

1 Line graph

Want to know

1 **The following bar graph represents Niigata City's average temperature per month. Let's look at the bar graph and discuss how the temperature changes and differs across months.**

Niigata City's average temperature per month

month	1	2	3	4	5	6	7	8	9	10	11	12
temperature (℃)	3	3	6	12	17	21	25	27	23	16	11	6

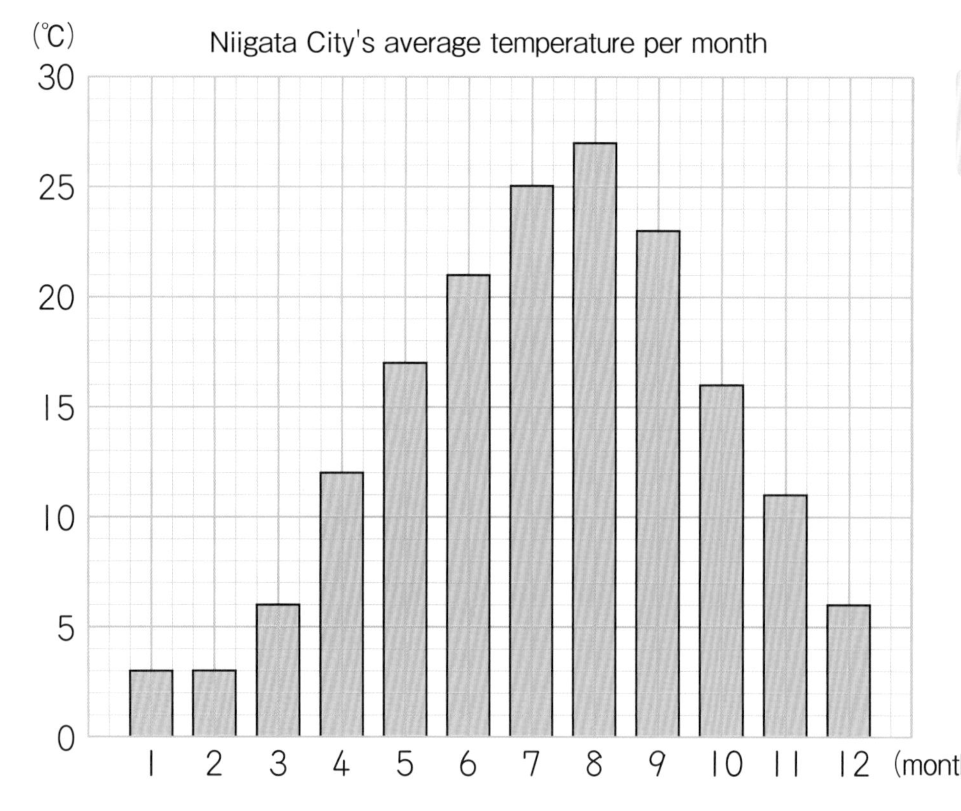

The unit to represent temperature is ℃.

As for graph representation, except for bar graphs, are there other options?

Nanami

① Draw a point at the end of the bar and connect them with a line as it is represented on the following graph. What are the horizontal and vertical axis representing?

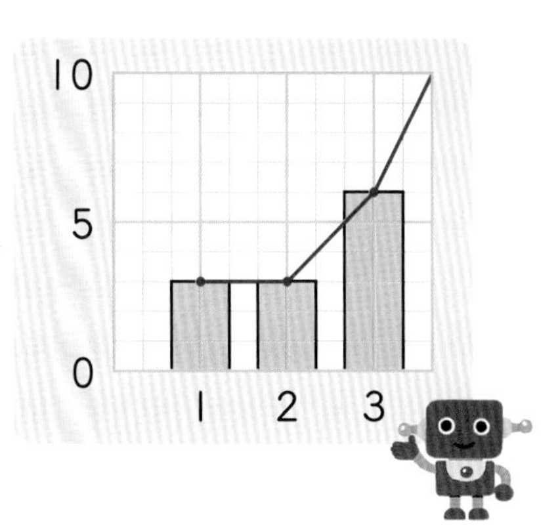

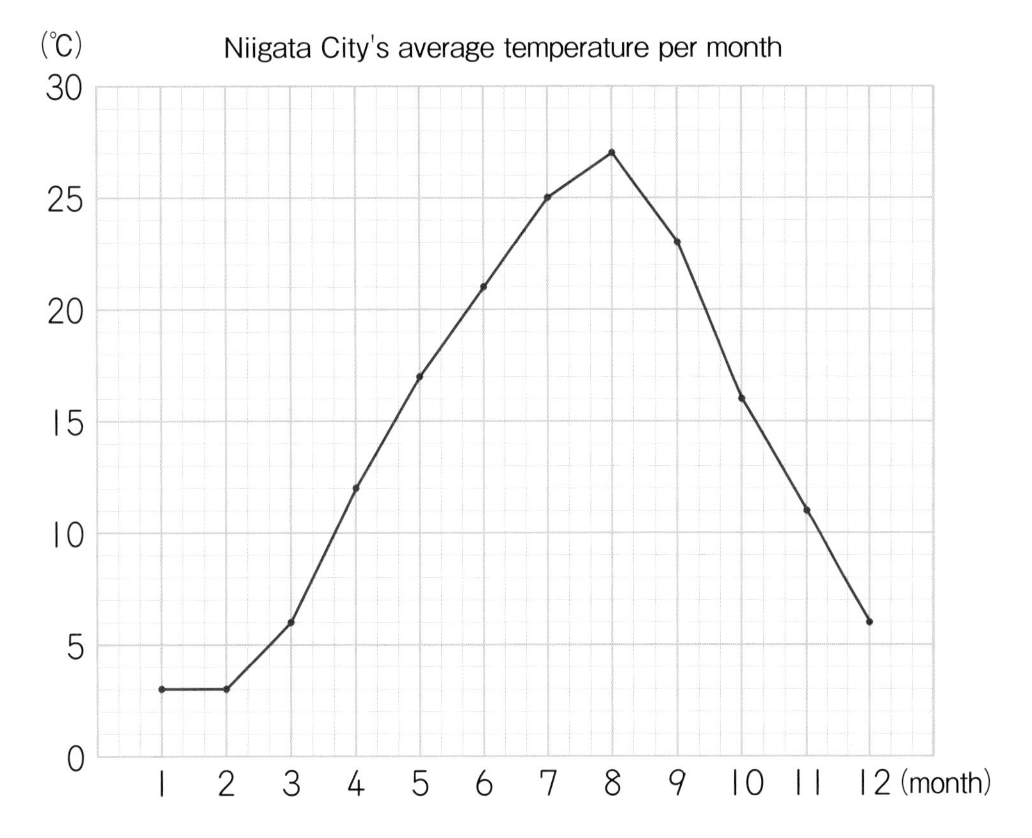

② What is the temperature (℃) in March? Also, in which month is the temperature 16℃?

Like above, a graph that represents the rate of change in temperature, etc., is called a **line graph.**

2 **The graph below represents Niigata City's and Auckland City's (New Zealand) average temperature per month. Let's compare how the temperature changes.**

Auckland City's average temperature per month

month	1	2	3	4	5	6	7	8	9	10	11	12
temperature (℃)	20	20	18	16	14	12	11	12	13	14	16	18

Niigata City's and Auckland City's average temperature per month

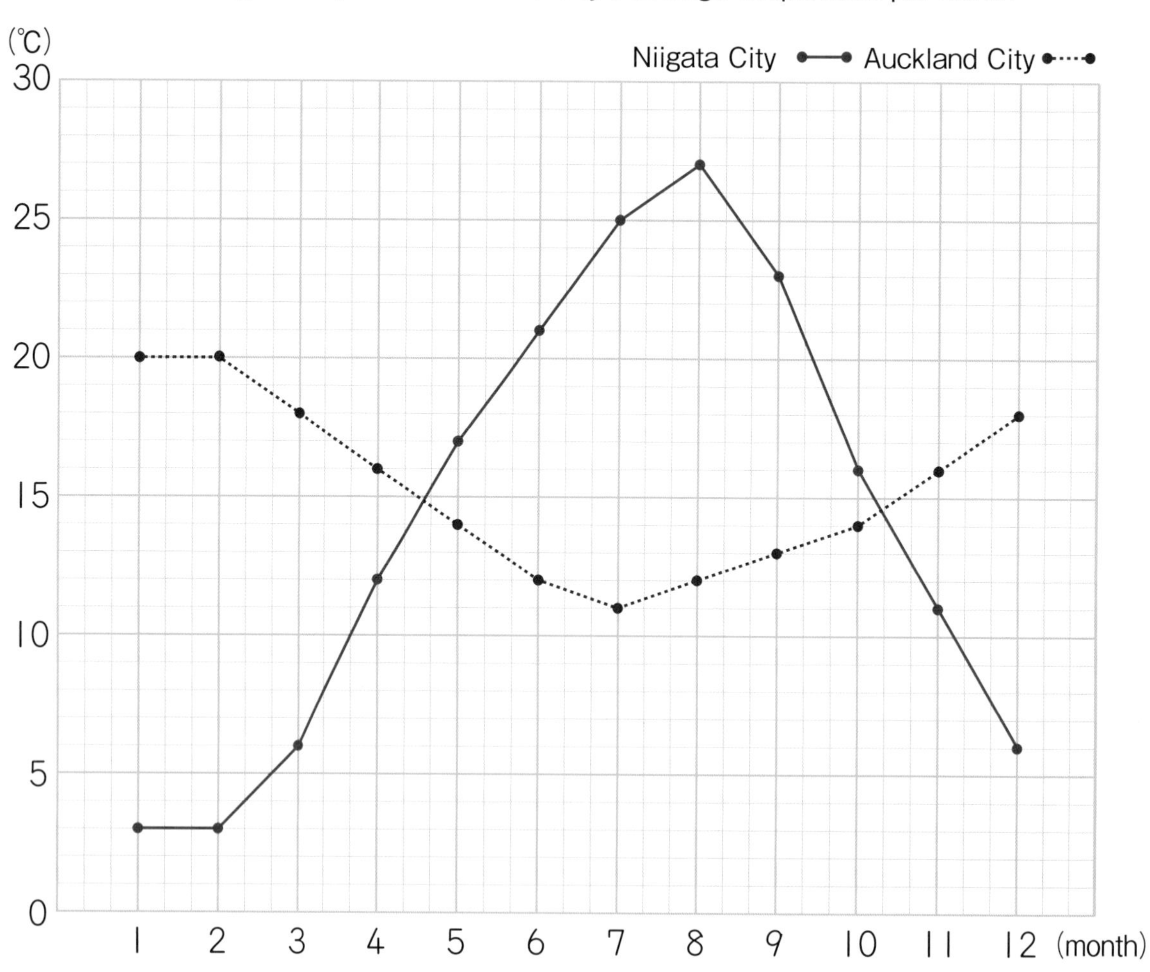

Want to think

① Which month has the highest temperature in each city? How many degrees are they?

② How does the temperature change? Let's compare how the temperature changes in Niigata City and Auckland City.

③ Which city has the highest change in temperature between two consecutive months? Name the respective months.

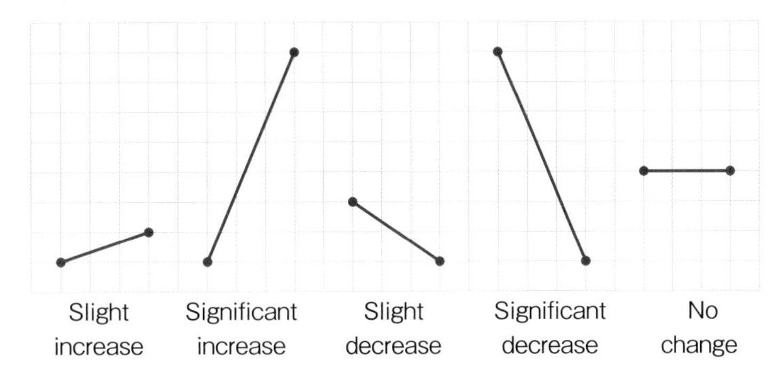

④ Let's discuss the benefits of line graphs.

We can easily compare the differences if we draw them in the same graph sheet.

Daiki

Want to confirm

Considering situations Ⓐ~Ⓕ, which is good to be represented by a line graph?

Ⓐ Your body temperature taken at the same time every day.

Ⓑ The type and number of vehicles that passed by your school in a period of 10 minutes.

Ⓒ The type of favorite fruit of your classmates and the number of the classmates for each type.

Ⓓ The temperature recorded every hour at the same place.

Ⓔ The height of your classmates.

Ⓕ Your height measured on your birthday every year.

2 How to draw a line graph

Want to know

1 **The table below shows how temperature changed in Yamaguchi City. Let's represent this information in a line graph.**

How temperature changed in Yamaguchi City (April 23rd, 2017)

time (hr)	a.m. 9	10	11	12	p.m. 1	2	3
temperature (℃)	15	17	18	20	21	22	22

How to draw a line graph

① Write units for the horizontal and vertical axis.

② Write time digits in the horizontal axis with the same spacing, and write a scale that represents 22 ℃ as the highest temperature in the vertical axis.

③ Based on the table, put a dot for each corresponding time and temperature.

④ Connect the dots with a line.

⑤ Write the title.

The table below shows how temperature changed in Naha City. Let's represent this information in a line graph.

Naha City's average temperature per month

month	1	2	3	4	5	6	7	8	9	10	11	12
temperature (℃)	17	17	19	21	24	27	29	29	28	25	22	19

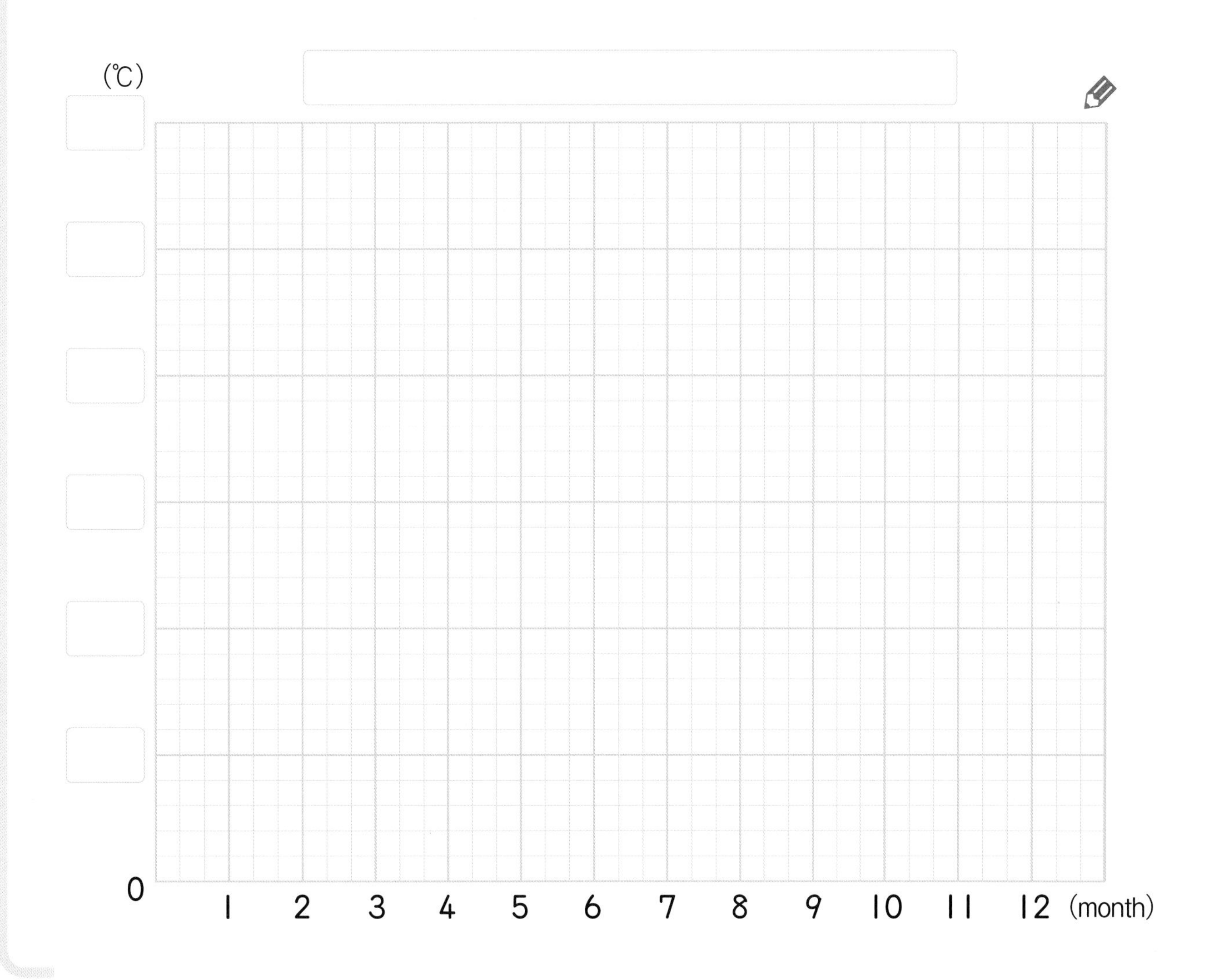

2 **The table below shows how the temperature changes near the window and corridor of Yumi's classroom. Let's think about the questions below.**

How the temperature changes near the window and corridor

time (hr)	a.m. 9	10	11	12	p.m. 1	2	3
Temperature near window (℃)	15	18	20	21	23	23	22
Temperature near corridor (℃)	15	16	18	18	20	20	19

① Let's represent in a line graph.

② Let's look at the graph and write what you understand.

Yui: It is easier to understand when both graph lines have different color and trace type.

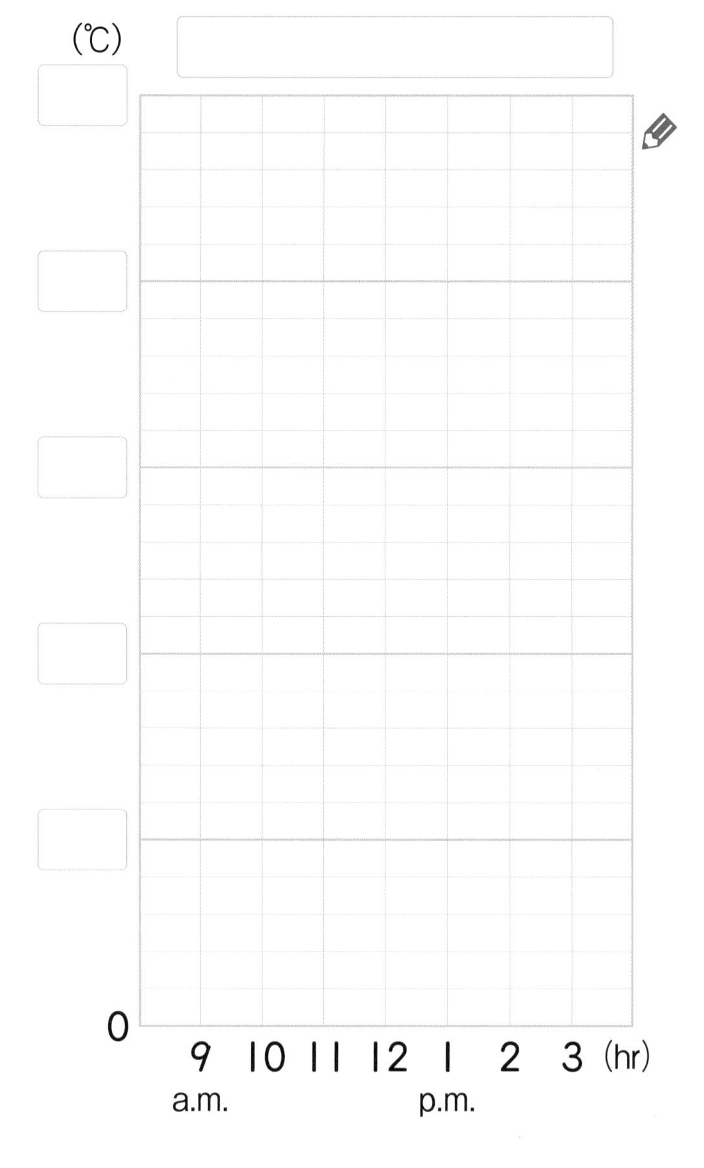

2 Ayumi investigated about how a shadow's length changed. The following tables show the records of the 10 cm stick's shadow in June and December. Let's represent the tables in a line graph.

Length of the shadow in June

time (hr)	8	9	10	11	12	13	14	15
shadow's length (cm)	12.1	7.9	4.9	2.8	2.1	3.5	6	9.3

Length of the shadow in December

time (hr)	8	9	10	11	12	13	14	15
shadow's length (cm)	51	27.8	20	16.8	16.3	18.1	23.1	36.1

① For each month, which is the biggest change in length between two consecutive hours? Name the hours.

② Let's look at the graph and write what you understand.

Why did the shadow's length become the shortest at noon?

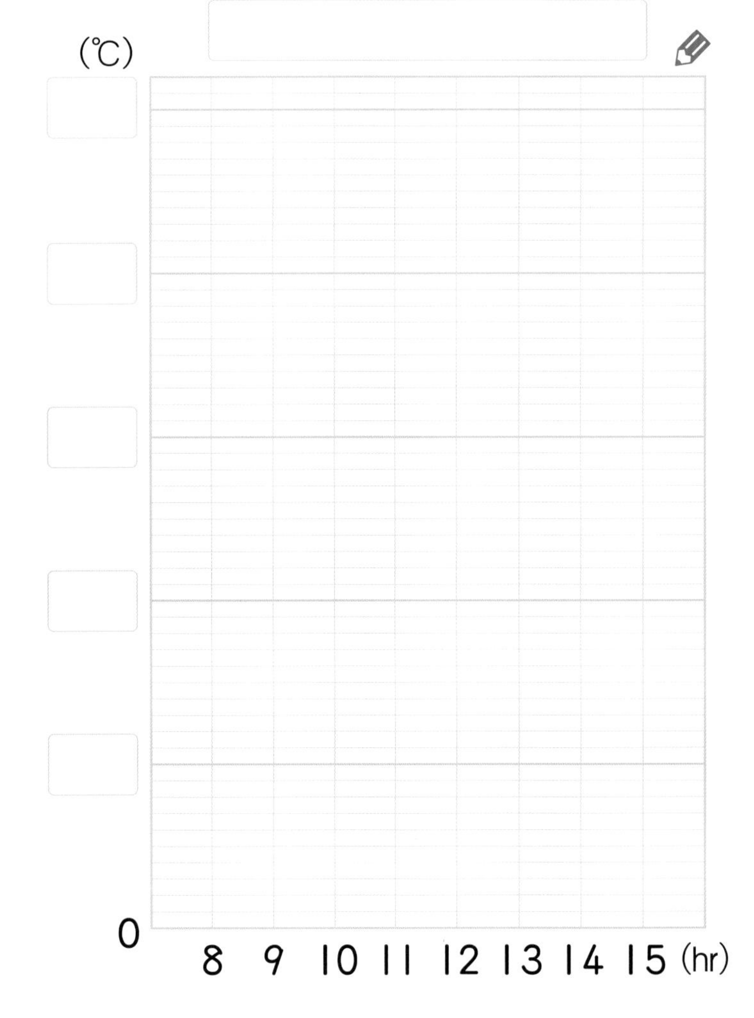

3 Ideas for drawing a line graph

Want to know

1 **When Yukie had a cold, she investigated how her body temperature changed and represented it in a line graph.**
Let's think about this graph.

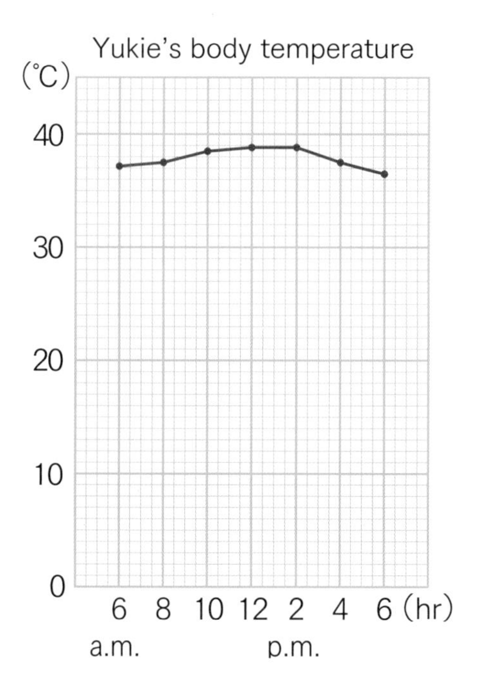

① What was her temperature (℃) at 8 a.m. and 2 p.m.?

Want to discuss

② In order to better understand how the body temperature changes, Yukie rewrote the following graph. Let's discuss her new idea.

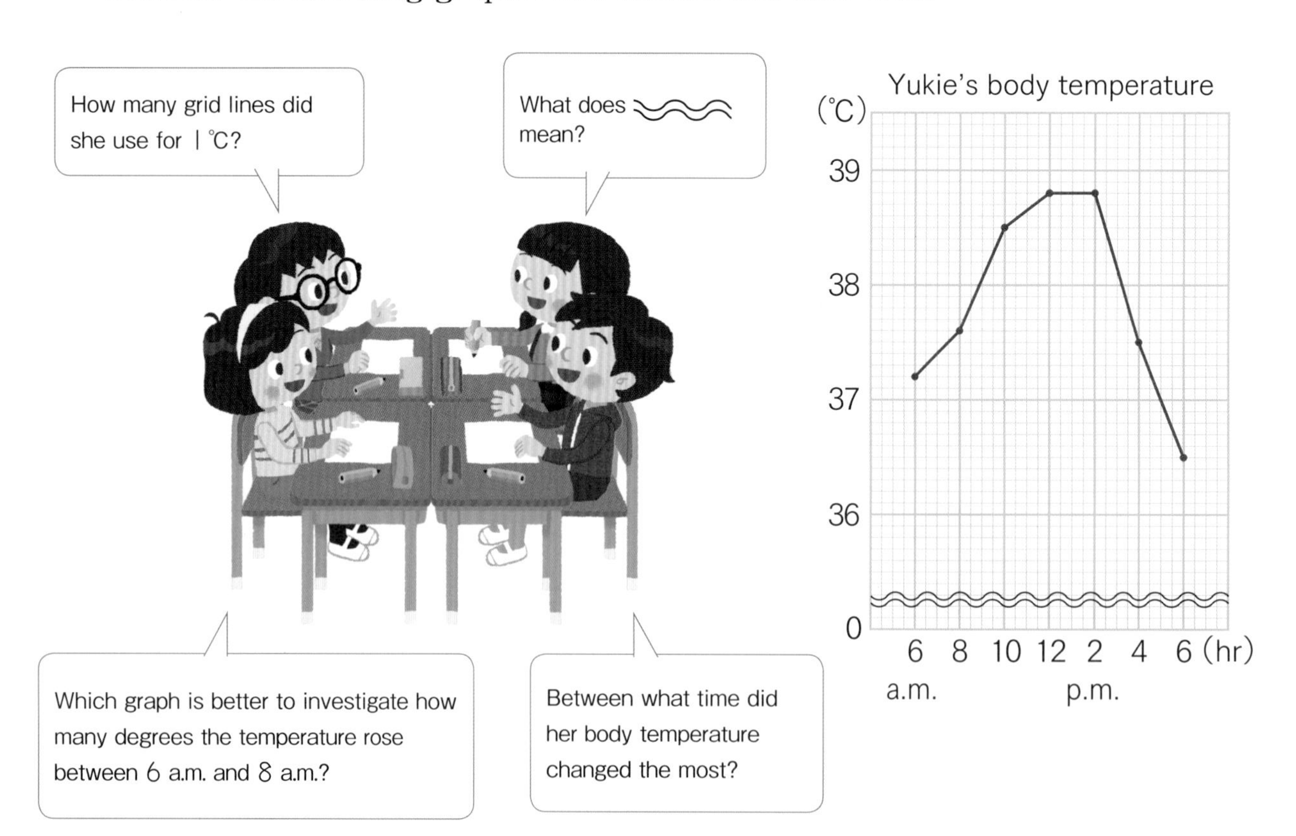

What you can do now

Understanding the rate of change from graphs.

1 The following table shows how the temperature changed. Let's answer the following questions.

How temperature changed

time (hr)	a.m. 9	10	11	12	p.m. 1	2	3	4	5
temperature (℃)	4	4	6	9	11	14	13	9	7

① Let's represent how the temperature changed in a line graph.

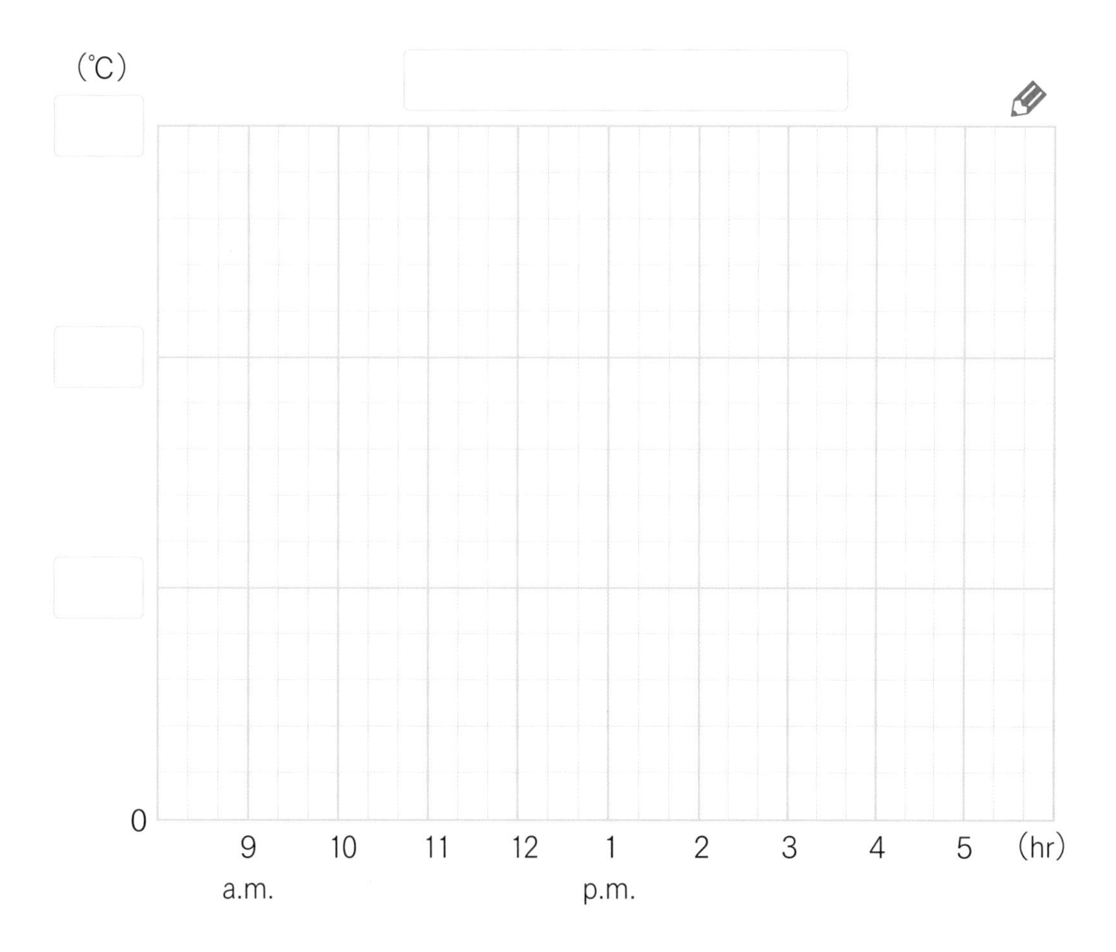

② Between what time did the temperature change the most?

③ Between what time did the temperature not change?

④ Let's look at the graph and write what you understand.

Supplementary problems p.150

Usefulness and efficiency of learning

1 The following table shows how Ryota's weight changed.
Based on this, the graph became as shown in ⓐ.
Let's answer the following questions.

Understanding the rate of change from graphs.

How weight changes

month	4	5	6	7	8	9	10	11
weight (kg)	27	27.5	28.4	28.7	29	29.2	29.5	30

ⓐ

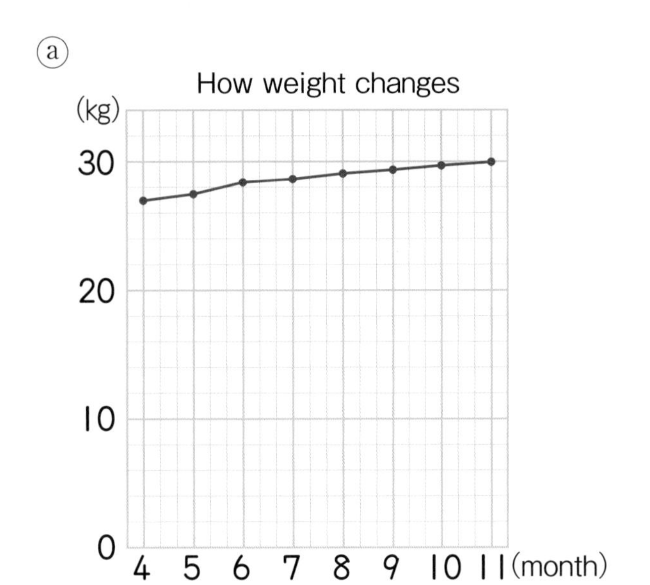

ⓑ

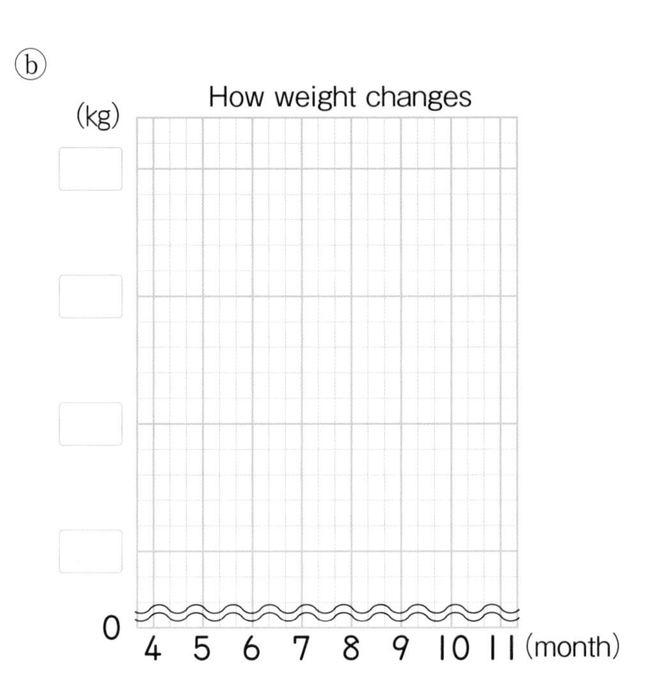

① In order to understand better the changes in weight, let's rewrite the graph on grid paper ⓑ. On the process, let's also write the vertical scale.

② Compare ⓐ and ⓑ graphs, where does ⓑ have a new idea?

③ Between which two consecutive months was the weight's way of increasing the highest?

④ Between which two consecutive months was the weight's way of increasing the lowest?

Become a writing master.

Notebook for thinking

Must write the ideas and doubts that you thought.

April 26

Write today's date.

Let's compare the line graphs that represent Niigata City's and Auckland City's average temperature per month. Let's write the things that you noticed.

Write the problem of the day that you must know.

When the temperature is high in one, the other is low. Why?

Write the doubts that you had.

〈My idea〉

○ In summer, the temperature is high in Niigata but low in Auckland.

○ Compared with Auckland, the monthly changes in Niigata are larger.

〈Classmate's idea〉

○ The largest increase in temperature was 6 ℃ from March to April in Niigata.

Write the classmate's ideas you consider good.

〈Reflect〉

○ With a line graph, we can easily understand the state of change.

○ I want to try to explore why the temperature changes in Niigata and in Auckland are different.

As for reflection, the following must be written.

- things you understood,
- interesting thoughts,
- things you noticed,
- and what you want to do next.

Reflect Connect

Problem

Is there any difference between the reading volume of 2 schools?

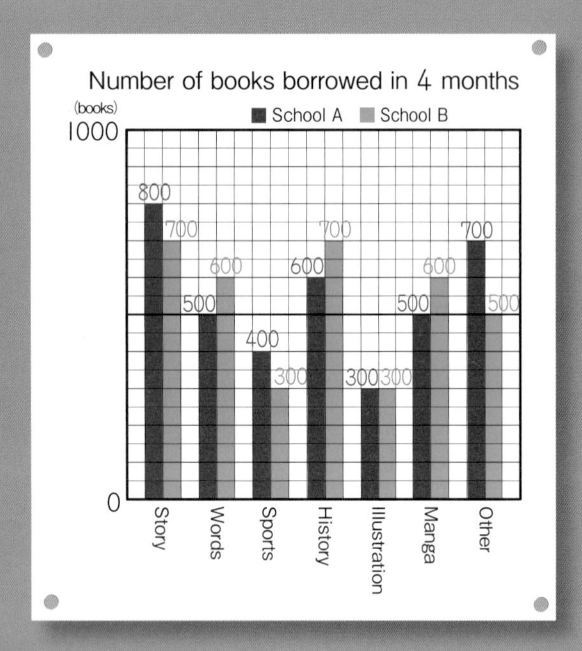

(books)

Book type	School A	School B	Total
Story	800	700	1500
Words	500	600	1100
Sports	400	300	700
History	600	700	1300
Illustration	300	300	600
Manga	500	600	1100
Other	700	500	1200
Total	3800	3700	7500

Let's write what you can read from the bar graph and table.

(Bar graph)

・As for School A, Story is the most borrowed type of book.

・As for School B, Story and History are the most borrowed type of book.

・As for School A and B, Illustration has the same number of borrowed books.

(Table)

・In a period of 4 months, School A borrowed more books.

The bar graph is easy to understand which type of book was borrowed the most.

The bar graph is difficult to understand how much is the increase.

In a line graph, increases and decreases are easy to understand.

Let's read change from the line graph.

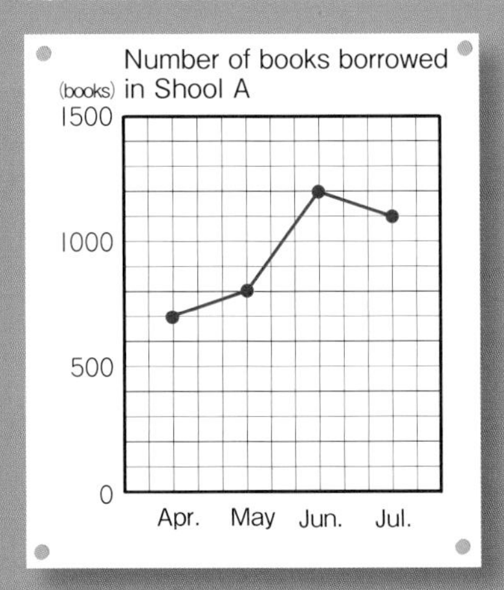

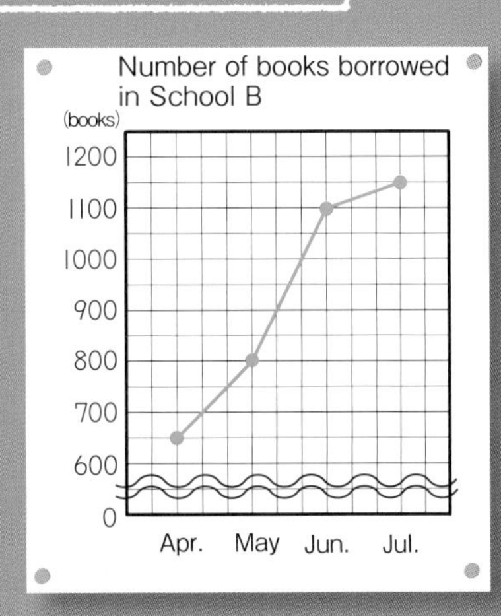

Hard to compare.

Let's have the same scale.

If the scale is aligned...

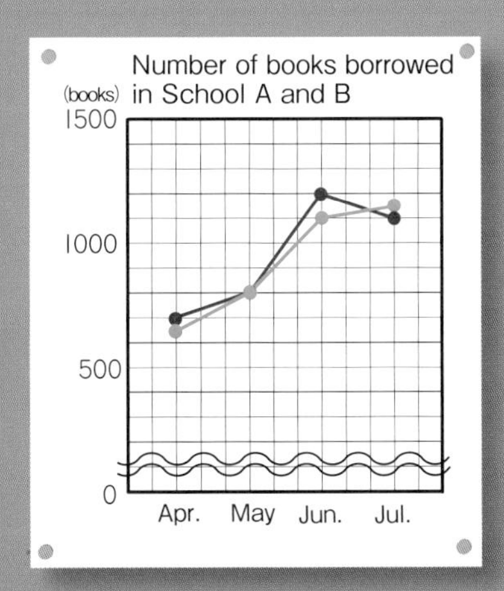

Things observed and understood from the line graphs.

- The highest increase was presented in School A between May and June.
- School A's lending amount decreased between June and July.
- School B's lending amount increased between April and July.

In a line graph, changes are easy to understand.

Want to connect

In line graphs, they don't have the same scale, it is difficult to compare.

Daiki

In this line graph, I cannot understand what type of book increased. I also want to try and create a line graph with types of books.

Nanami

2 digit-number division?

Problem Let's think about how to calculate $48 \div 3$.

The case: (2-digit) ÷ (1-digit)

Let's think about how to calculate.

Want to solve

Activity

48 sheets of colored paper are divided equally among 3 children.

How many sheets of paper will receive one child?

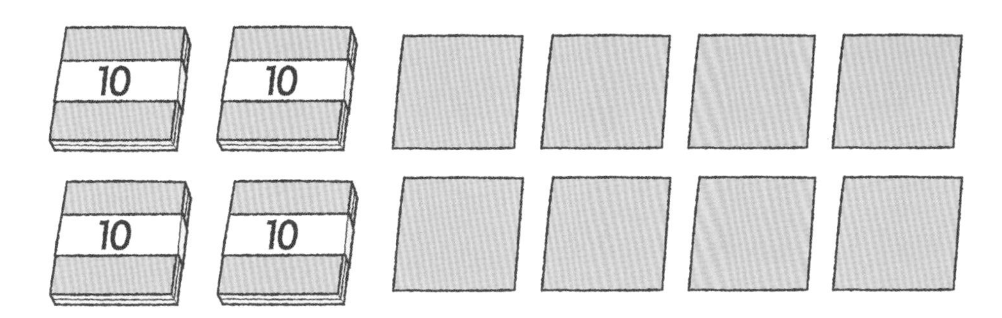

① Let's write a math expression.

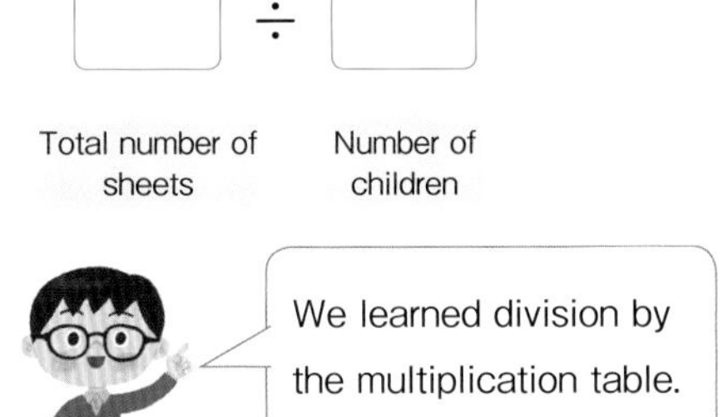

☐ ÷ ☐

Total number of sheets　Number of children

Hiroto: We learned division by the multiplication table.

Nanami: With this division, we cannot use the multiplication table.

② About how many sheets will it become?

Daiki: Is the answer a number larger than 10?

Want to think

③ Let's think about how to calculate, using what you have learned so far.

Let's think about various calculation methods and explain using diagrams and math expressions.

Want to explain

④ Let's explain the ideas of the following 3 children.

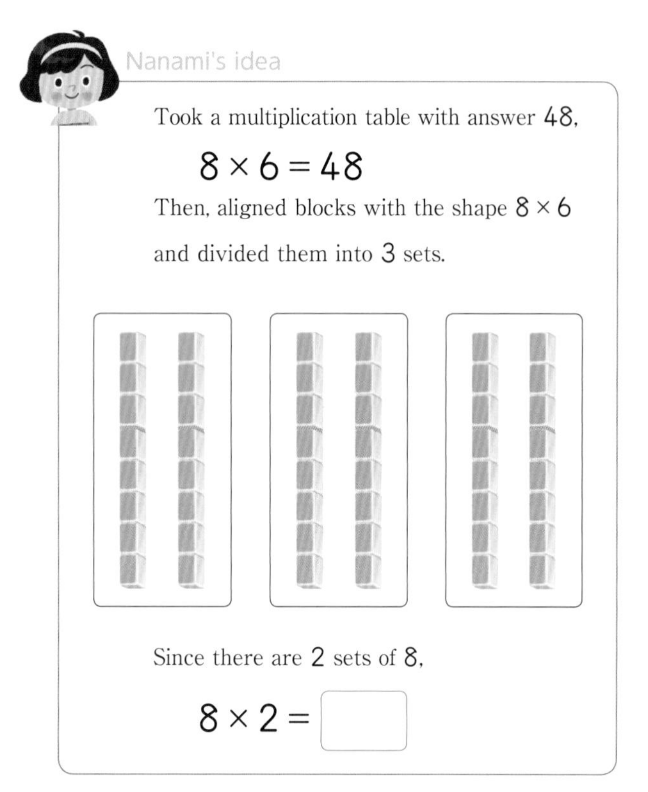

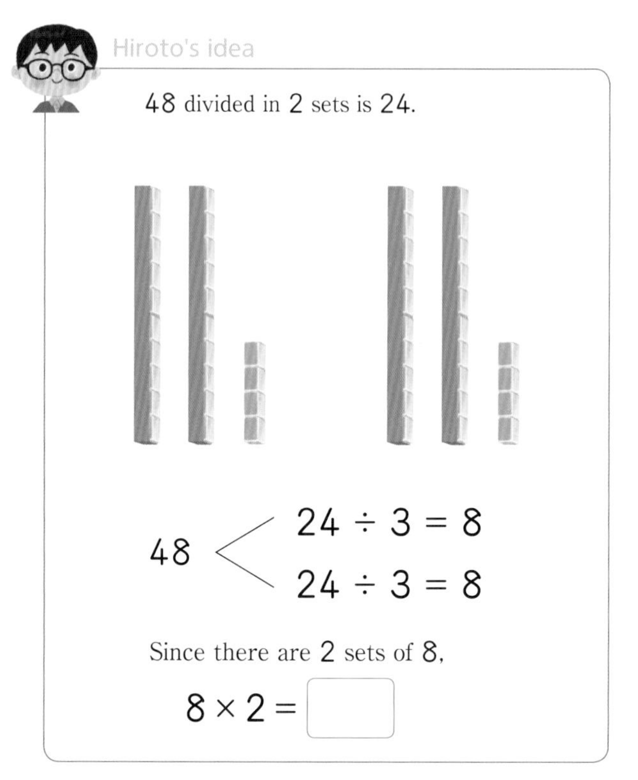

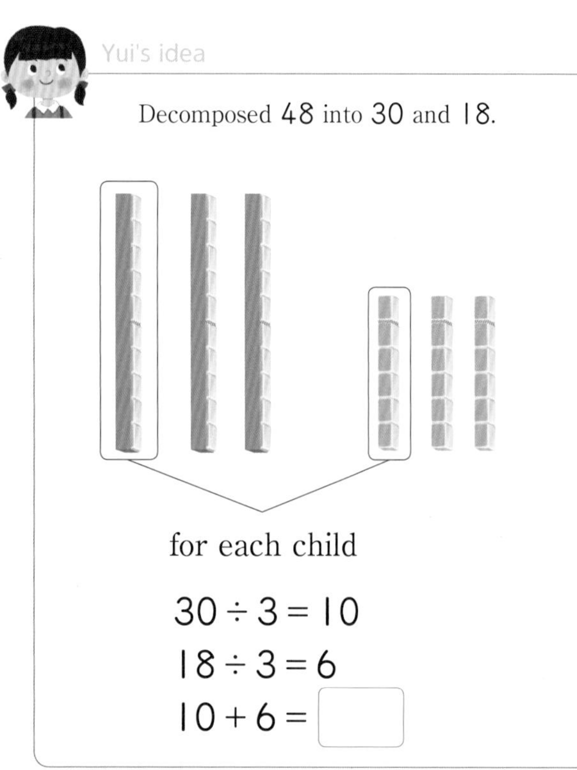

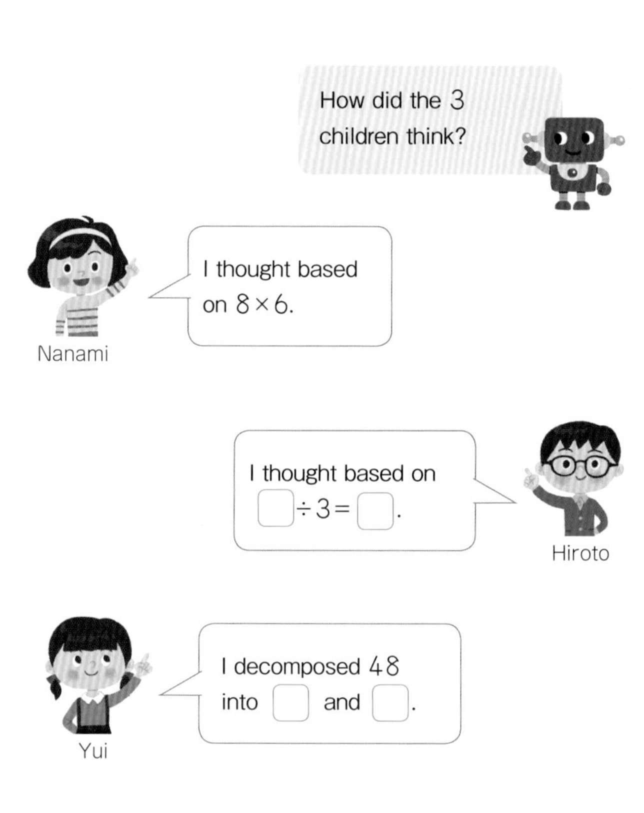

Want to compare

⑤ Let's compare your own and friend's idea.

⑥ Let's summarize what you have learned in your notebook.

Way to see and think

Where are your own and friend's ideas similar?

Want to represent

Let's summarize the calculation method for 56 ÷ 4 in your notebook.

How to calculate 56 ÷ 4

1 : How to think

- First, separate 4 sets of 10.
- Then, divide the remaining by 4.

2 : How to solve

(Diagram)

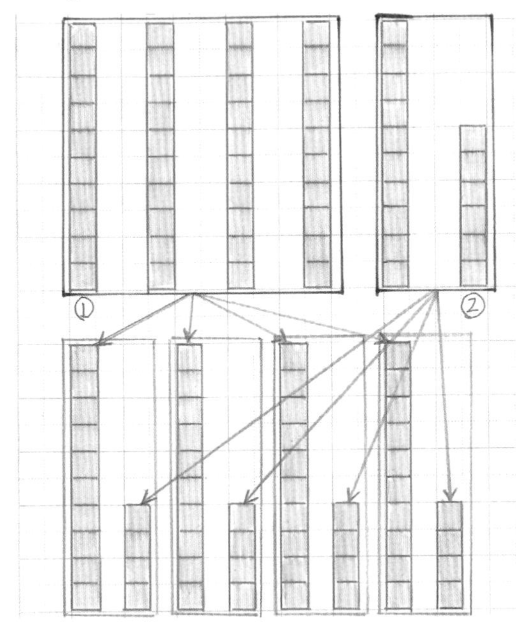

(Math sentence)

① 40 ÷ 4 = 10

② 16 ÷ 4 = 4

Altogether

10 + 4 = 14

Answer : 14

3 : Understood things

Even when the dividend is a large number, we can solve by using what we have learned until now.

Decompose the dividend into two.

Write the title.

[1: How to think]
Write what you thought to solve the problem.

[2: How to solve]
Write how you solved the problem using diagrams, math expressions, words, etc., in order to understand easily.

[3: Understood things]
Write what you understood and found out.

Division by 1-digit number

Let's think about the division in vertical form.

1 Division with 1-digit quotient

Want to know: Division in vertical form

1 **We want to divide 48 sheets of colored paper such that 9 children receive the same number. How many sheets of paper will each child receive and how many sheets of paper will remain?**

① Let's write a math expression and find the answer.

☐ ÷ ☐

Total number of sheets / Number of children

Answer: Each child receives ☐ sheets of paper and the remainder is ☐ sheets of paper.

② Let's solve $48 \div 9$ in vertical form.

Division algorithm to calculate $48 \div 9$ in vertical form

Write and calculate as shown on the right.

① Write 5 above the ones place of 48.

② Write 45, from "9 times 5 is 45," in the aligned place below 48.

③ Subtract 45 from 48. The remainder is 3.

④ Confirm that the remainder 3 is smaller than the divisor 9.

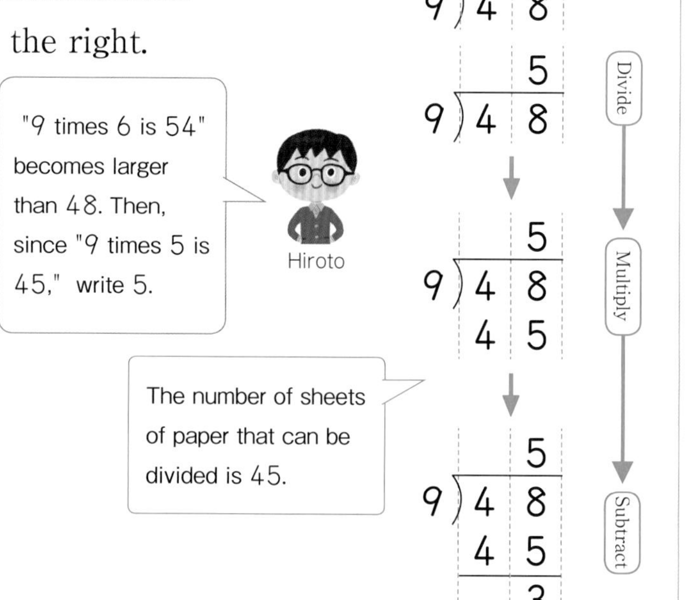

As addition, subtraction, and multiplication, division can also be solved in vertical form.

Want to know

2 Let's think about the division algorithm to calculate $48 \div 8$ in vertical form.

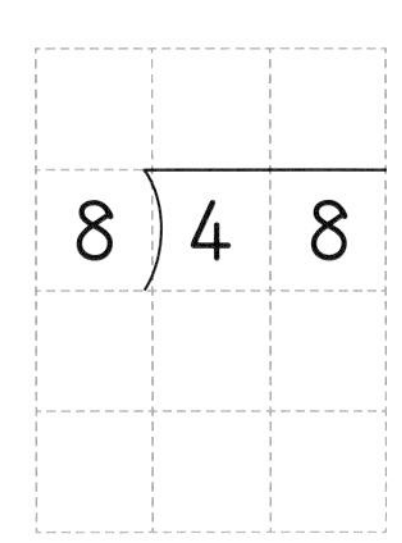

8)4 8 Way of writing

① 4 8

②)4 8

③)4 8 (with the line drawn over 4 8)

④ 8)4 8

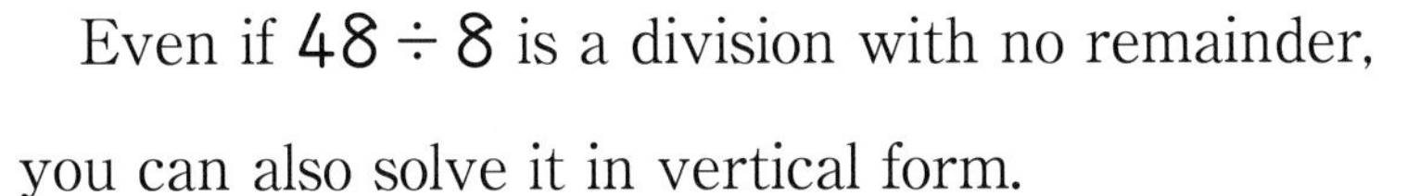

Even if $48 \div 8$ is a division with no remainder, you can also solve it in vertical form.

The answer for divisions with remainder becomes a **quotient** and a **remainder**.

Want to confirm

1 Let's confirm the answer of the following divisions.

① $48 \div 8 = 6$

Dividend, Divisor, Quotient

$8 \times 6 = \square$

Divisor, Quotient, Dividend

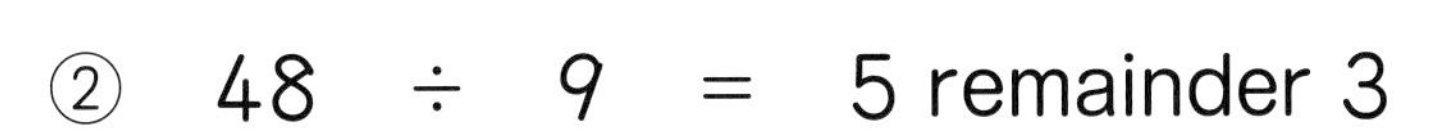

② $48 \div 9 = 5$ remainder 3

Dividend, Divisor, Quotient, Remainder

$9 \times 5 + 3 = \square$

Divisor, Quotient, Remainder, Dividend

Want to try

2 Let's solve the following calculations in vertical form. Also, let's confirm the answer.

① $37 \div 5$　② $56 \div 9$　③ $51 \div 7$　④ $65 \div 8$

⑤ $13 \div 2$　⑥ $7 \div 3$　⑦ $21 \div 7$　⑧ $8 \div 2$

2 Division of tens and hundreds

Want to solve

1 **We want to divide 60 sheets of colored paper such that 3 children receive the same number. How many sheets of paper will each child receive?**

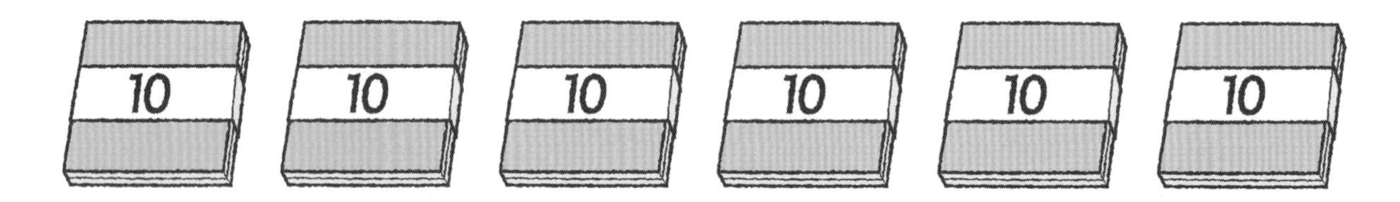

① Let's write a math expression.

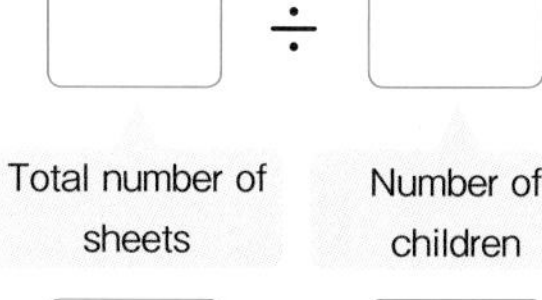

② If you decompose the total number into 6 bundles of 10 sheets and divide by 3 children, how many bundles will each child receive?

□ ÷ □

Number of 10 sheets bundles　Number of children

③ How many sheets of paper will each child receive?

Want to try

1 We want to divide 600 sheets of colored paper such that 3 children receive the same number. How many sheets of paper will each child receive?

① Let's write a math expression. □

② How many sheets of paper will each child receive?

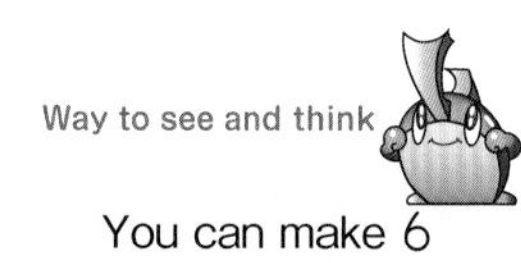

You can make 6 bundles of 100 sheets.

Want to confirm

2 Let's solve the following calculations.

① $80 \div 4$　② $90 \div 3$　③ $150 \div 5$

④ $300 \div 6$　⑤ $800 \div 2$　⑥ $1600 \div 4$

3 Division with 2-digit quotient

Want to solve

1 **We want to divide 69 sheets of colored paper such that 3 children receive the same number. How many sheets of paper will each child receive?**

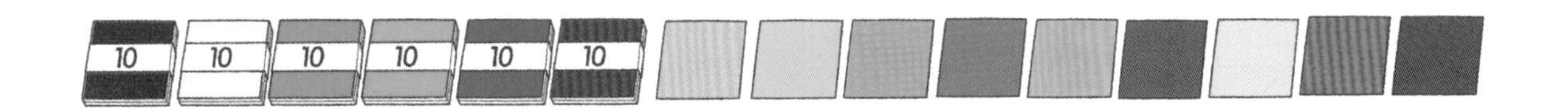

① Let's write a math expression.

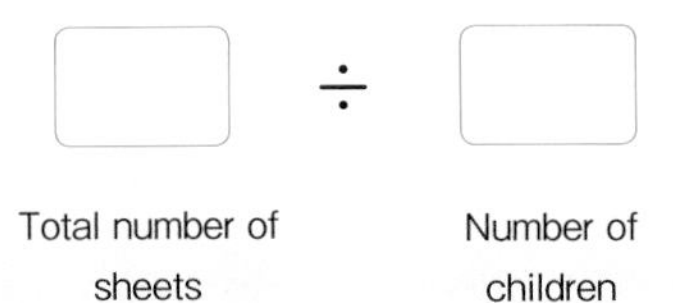

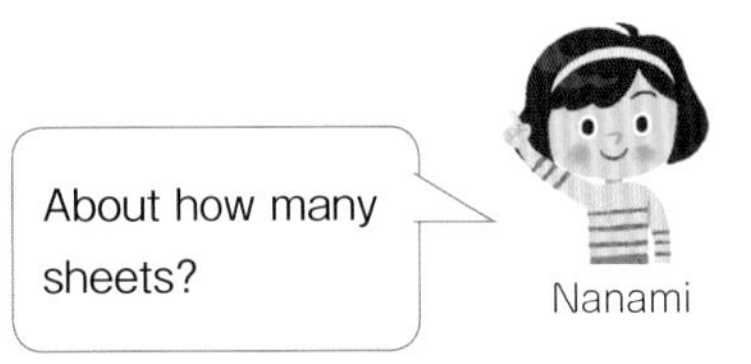

② Let's think about how to find the quotient of $69 \div 3$ by looking at the diagram on the right.

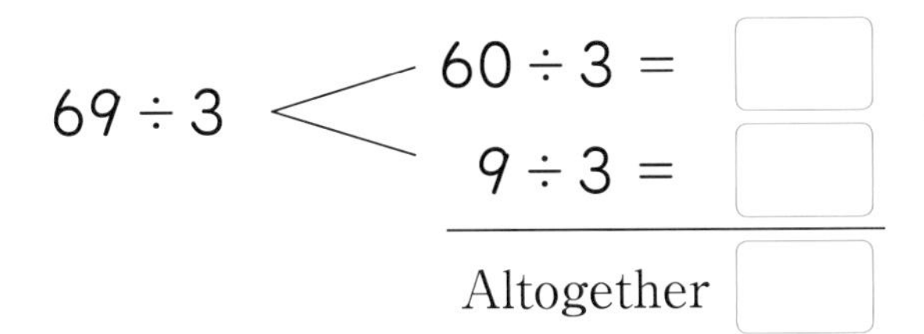

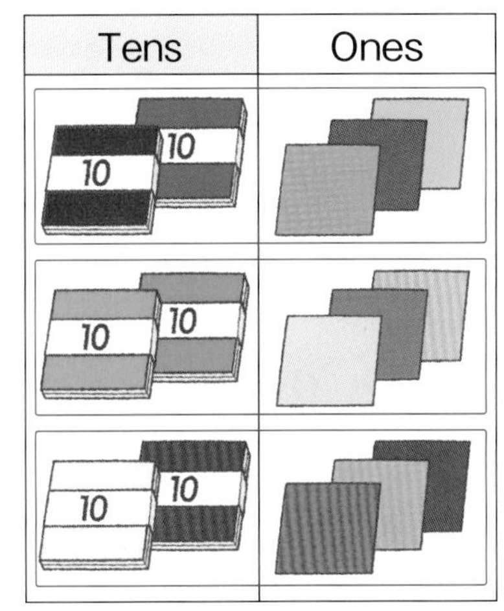

Want to try

We want to divide 72 sheets of colored paper such that 3 children receive the same number. How many sheets of paper will each child receive?

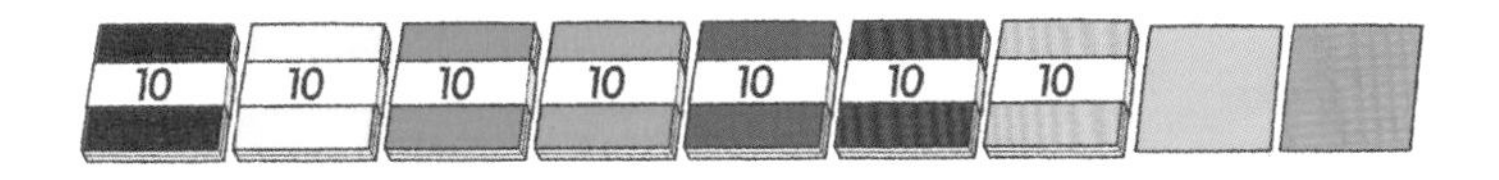

① Let's write a math expression.

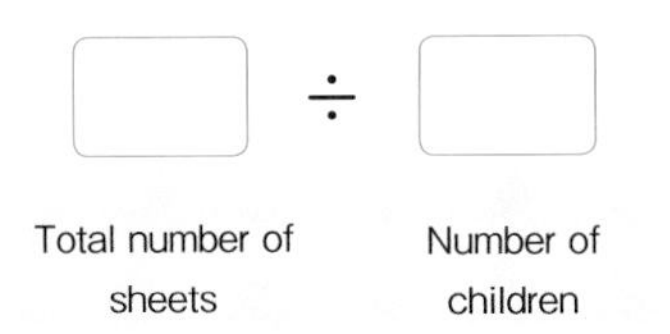

If I decompose bundles of 10 sheets, I will get a remainder.

② Let's think about how to calculate.

How to think $72 \div 3$

(1) If the 7 bundles of 10 sheets are divided among 3 children, how many bundles will each child receive and how many bundles will remain?

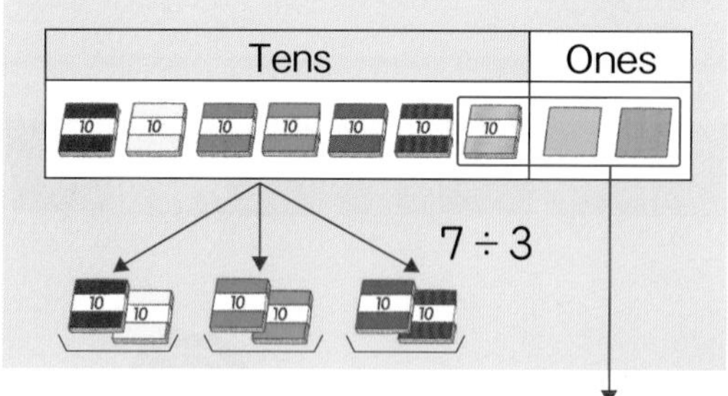

(2) Break up the remaining 1 bundle and combine that with the 2 single sheets.

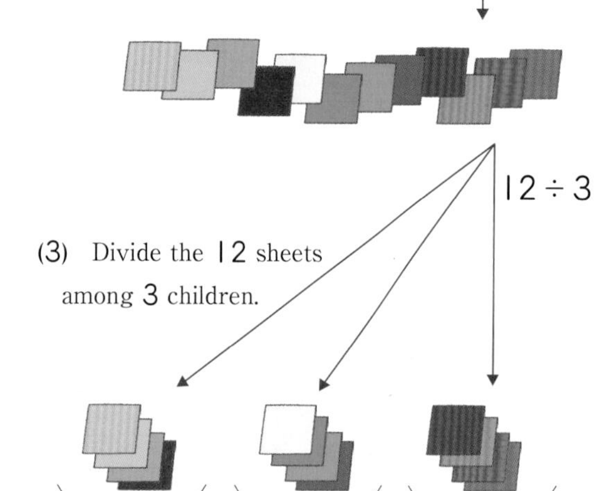

(3) Divide the 12 sheets among 3 children.

(4) As for each child,

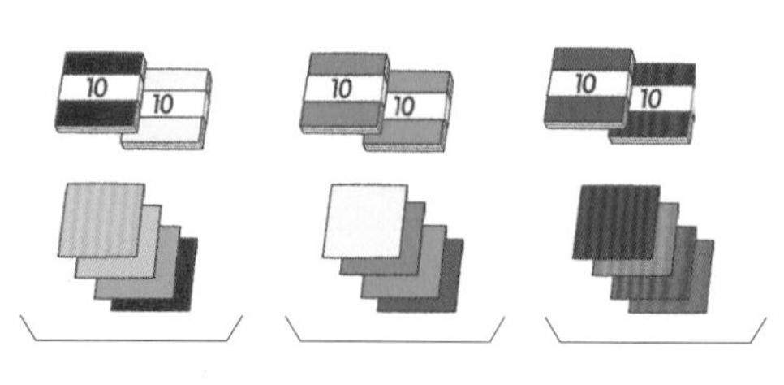

$72 \div 3$
$60 \div 3 =$ ☐
$12 \div 3 =$ ☐
Altogether ☐

Division algorithm to calculate $72 \div 3$ in vertical form

Tens place calculation

Divide

$7 \div 3 = 2$ remainder 1

Write 2 in the tens place.

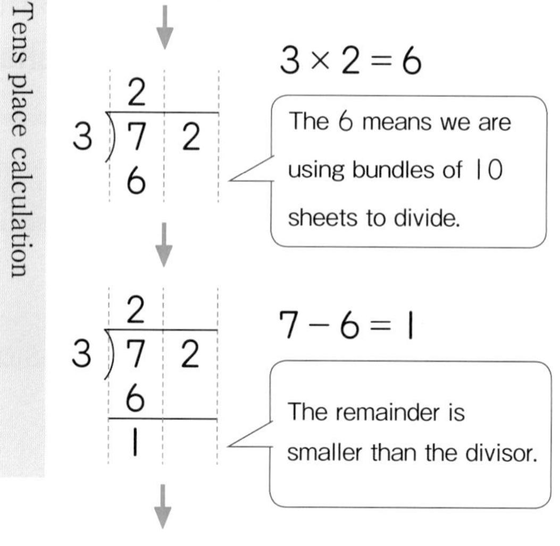

Multiply

$3 \times 2 = 6$

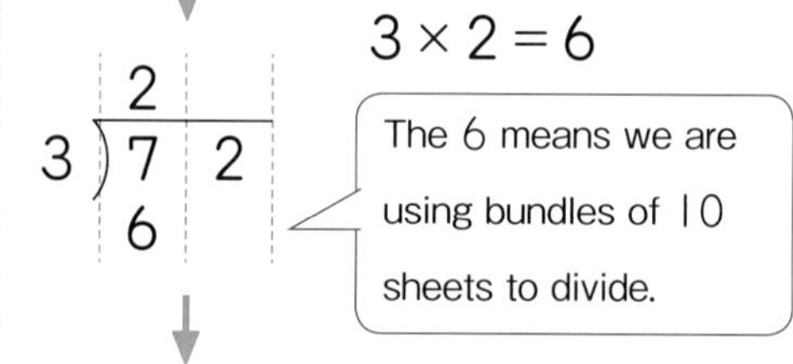

Subtract

$7 - 6 = 1$

The remainder is smaller than the divisor.

Bring down

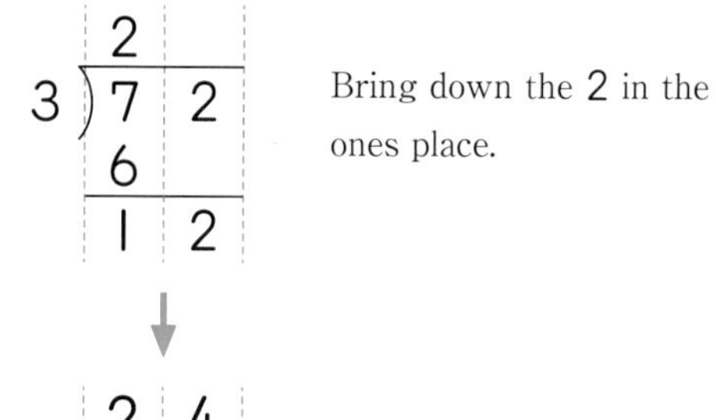

Bring down the 2 in the ones place.

Ones place calculation

Divide

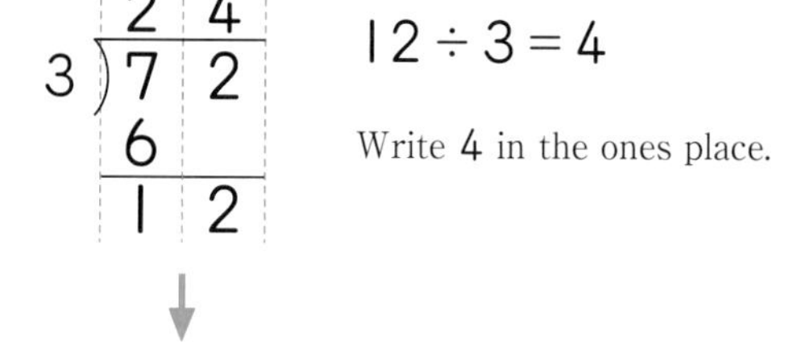

$12 \div 3 = 4$

Write 4 in the ones place.

Multiply

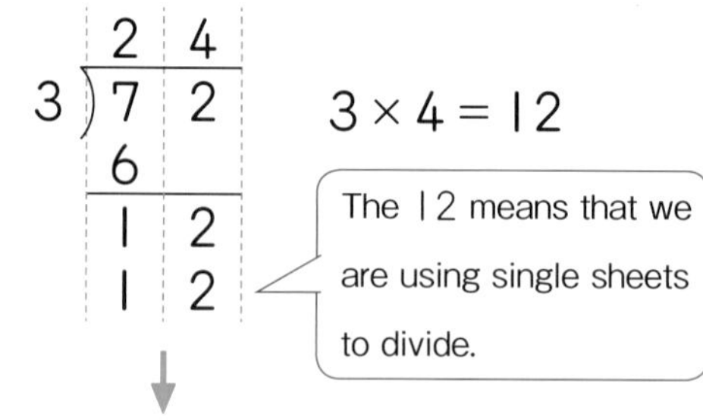

$3 \times 4 = 12$

Subtract

$12 - 12 = 0$

③ Let's solve $72 \div 3$ in vertical form.

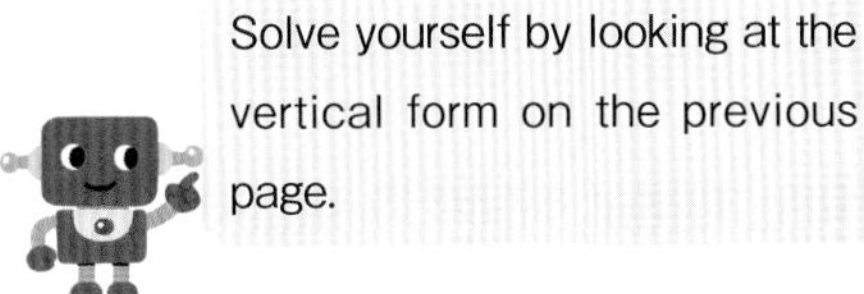

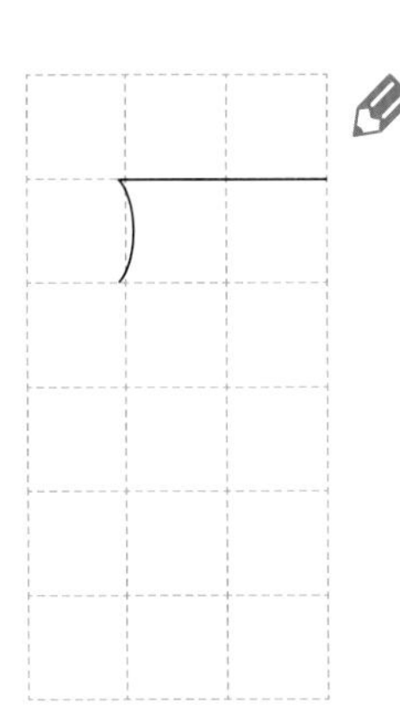

When solving divisions in vertical form, start in order from the highest place value.

Want to think

2 Koichi was solving $92 \div 4$ in vertical form. In the process he was in trouble. Let's think why he was in trouble, and calculate correctly.

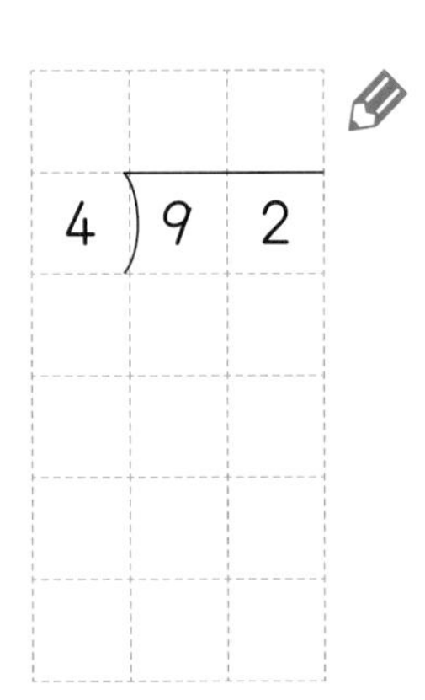

Want to confirm

Let's solve the following calculations in vertical form.

① $54 \div 2$ ② $68 \div 4$ ③ $34 \div 2$ ④ $84 \div 3$

3 **We want to divide 83 sheets of colored paper such that 5 children receive the same number. How many sheets of paper will each child receive and how many sheets of paper will remain?**

① Let's write a math expression. ☐

② Let's explain the division algorithm in vertical form.

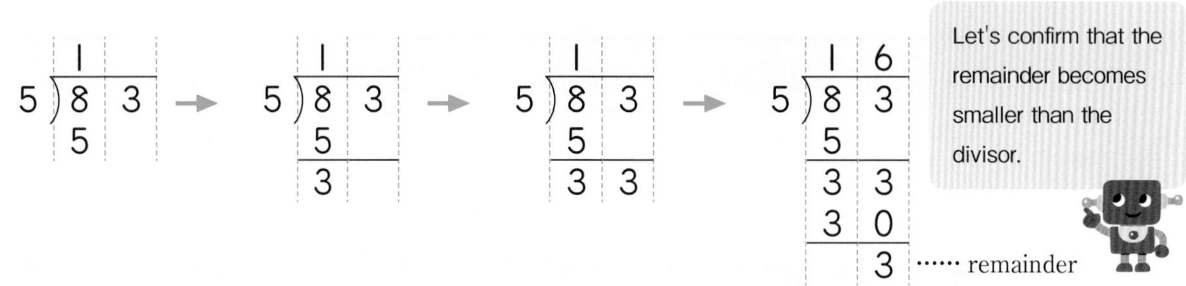

$8 \div 5 = 1$ remainder 3
Write 1 in the tens place.
$5 \times 1 = 5$

$8 - 5 = 3$

Bring down the 3 in the ones place.

$33 \div 5 = 6$ remainder 3
Write 6 in the ones place.
$5 \times 6 = 30$
$33 - 30 = 3$

$83 \div 5 =$ ☐ remainder ☐

Answer: Each child receives ☐ sheets of paper and the remainder is ☐ sheets of paper.

③ Let's confirm the answer.

5 × ☐ + ☐ = 83

Divisor × Quotient + Remainder = Dividend

Want to confirm

3 Let's solve the following calculations in vertical form. Also, let's confirm the answer.

① $37 \div 2$ ② $58 \div 4$ ③ $86 \div 7$

④ $93 \div 8$ ⑤ $76 \div 6$ ⑥ $47 \div 3$

Want to try

4 There is a 67 cm long tape. How many 5 cm long pieces can you make? How many cm is the remainder?

Let's explain the following division algorithm in vertical form.

① $64 \div 3$

```
   2            2            2 1
3)6 4   →   3)6 4   →   3)6 4
              6            6
              0 4            4
                             3
                             1
```

```
   2
3)6 4
   6
   0 4
```

There is no problem if you do not write the 0.

② $92 \div 3$

```
   3            3 0          3 0
3)9 2   →   3)9 2   →   3)9 2
              9            9
                2            2
                             0
                             2
```

Want to confirm

Let's solve the following calculations in vertical form.

① $45 \div 2$　② $97 \div 3$　③ $57 \div 5$

④ $62 \div 3$　⑤ $81 \div 2$　⑥ $82 \div 4$

Want to try

The division algorithms in vertical form on the right are wrong. Let's explain the reasons and solve correctly.

①

```
   2 7
3)8 5
   6
   2 5
   2 1
     4
```

②

```
   2
2)4 1
   4
     1
```

4 The case: (3-digit) ÷ (1-digit)

Want to solve

1 **There are 639 sheets of colored paper. If the papers are divided such that 3 groups receive the same number of sheets, how many sheets of paper will each group receive?**

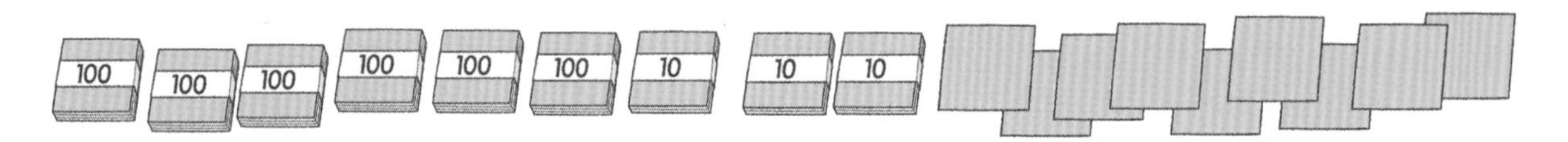

① Let's write a math expression.

② About how many hundred sheets of paper? $639 \div 3$

③ Let's think about the division algorithm in vertical form.

$600 \div 3 =$ ☐

$30 \div 3 =$ ☐

$9 \div 3 =$ ☐

Altogether ☐

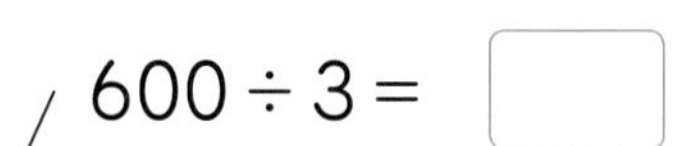

Want to try

1 We want to divide 536 sheets of colored paper such that 4 children receive the same number. How many sheets of paper will each child receive? Let's think about how to calculate.

$536 \div 4$

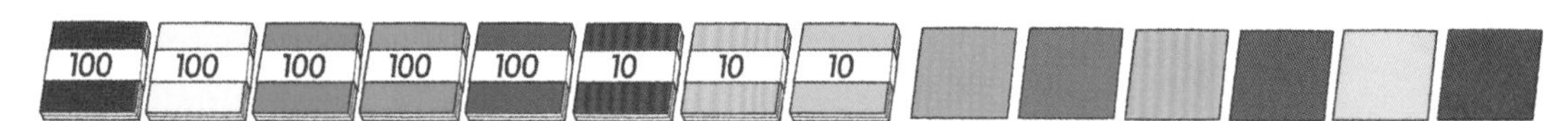

① Let's divide the bundles of 100 sheets.

$5 \div 4 =$ ☐ remainder ☐

Number of bundles of 100 sheets

② Let's divide the bundles of 10 sheets. ☐ $\div 4 =$ ☐ remainder ☐

③ Let's divide the single sheets. ☐ $\div 4 =$ ☐

④ How many sheets of paper will each child receive? $536 \div 4 =$ ☐

⑤ Let's think about the division algorithm in vertical form.

Division algorithm to calculate 536 ÷ 4 in vertical form

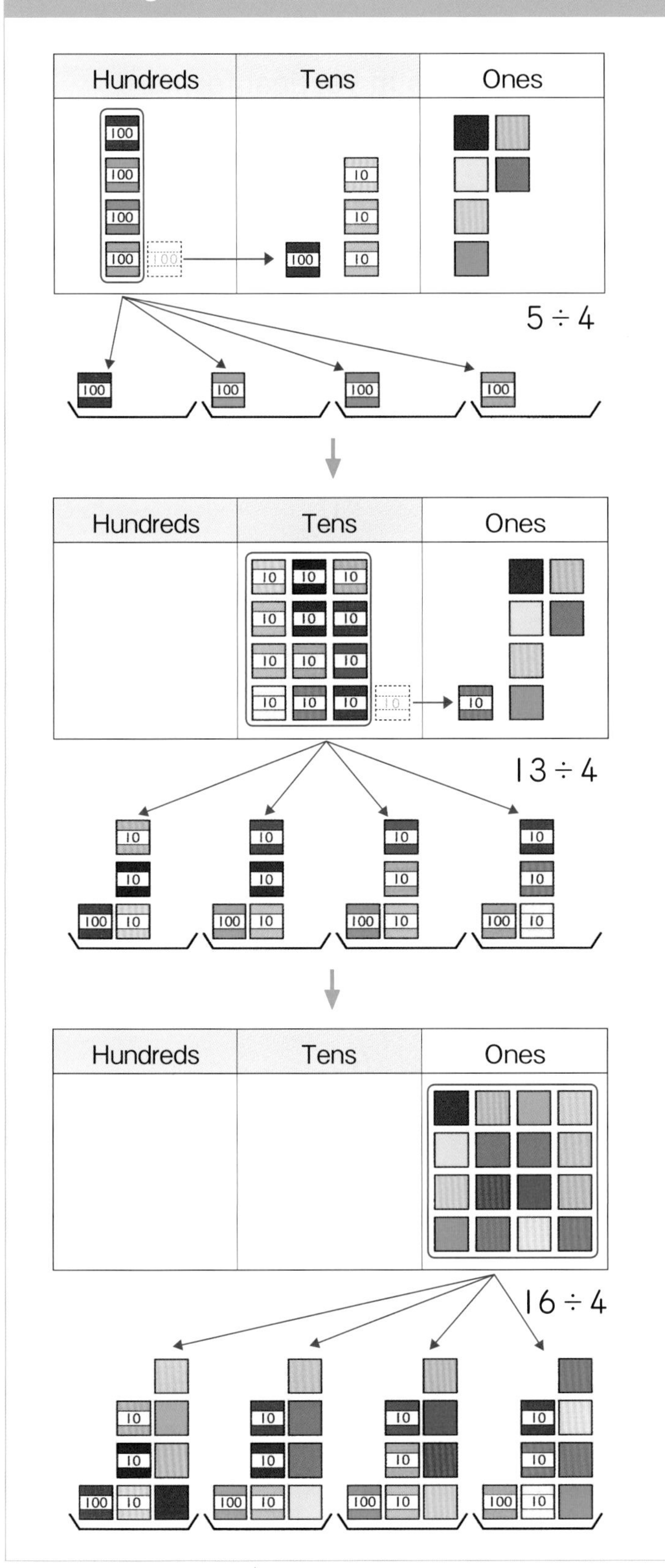

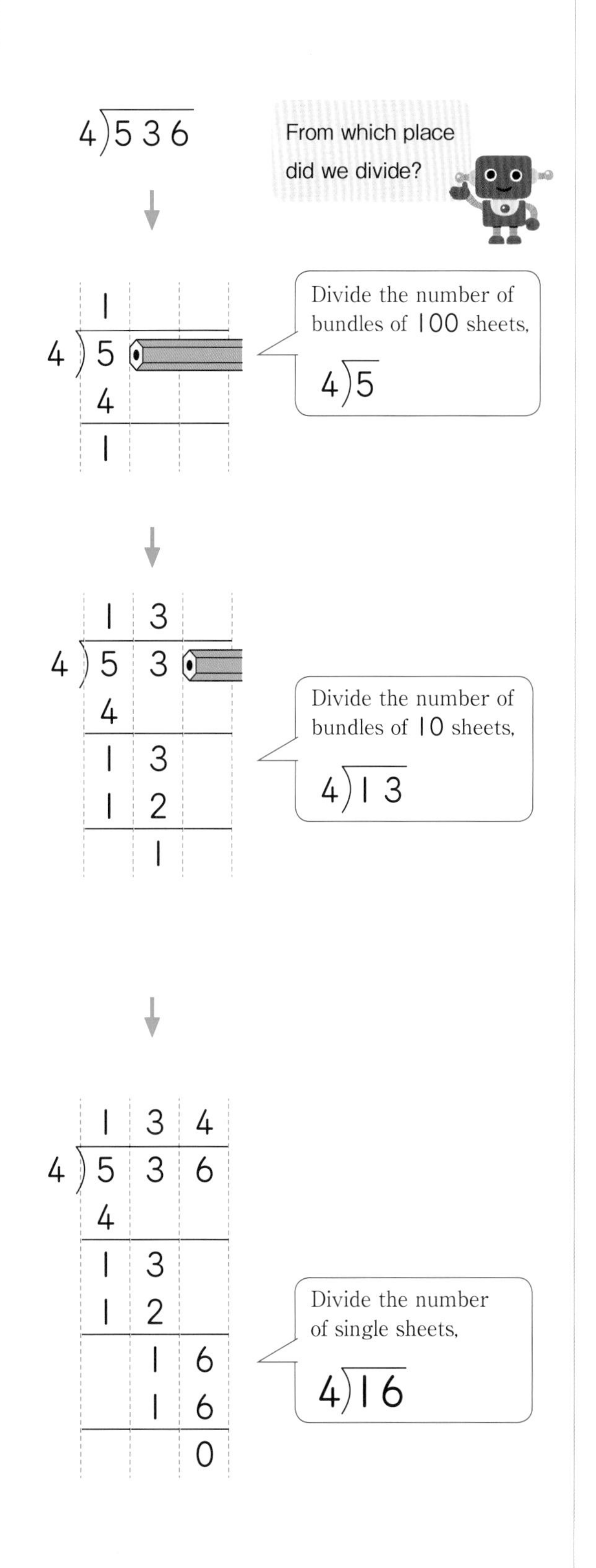

Want to explain Treatment of 0

2 Let's explain the following Ⓐ and Ⓑ divisions in vertical forms. Also, let's confirm the answers.

① $420 \div 3$

Ⓐ
$$\begin{array}{r} 140 \\ 3\overline{)420} \\ \underline{3} \\ 12 \\ \underline{12} \\ 0 \\ \underline{0} \\ 0 \end{array}$$

Ⓑ
$$\begin{array}{r} 140 \\ 3\overline{)420} \\ \underline{3} \\ 12 \\ \underline{12} \\ 0 \end{array}$$

② $859 \div 8$

Ⓐ
$$\begin{array}{r} 107 \\ 8\overline{)859} \\ \underline{8} \\ 5 \\ \underline{0} \\ 59 \\ \underline{56} \\ 3 \end{array}$$

Ⓑ
$$\begin{array}{r} 107 \\ 8\overline{)859} \\ \underline{8} \\ 59 \\ \underline{56} \\ 3 \end{array}$$

Want to confirm

Let's solve the following calculations in vertical form.

① $740 \div 2$　② $650 \div 5$　③ $742 \div 7$　④ $958 \div 9$

Want to try

Let's explain the mistake in the division algorithm in vertical form shown on the right. Calculate correctly.

$$\begin{array}{r} 27 \\ 3\overline{)622} \\ \underline{6} \\ 22 \\ \underline{21} \\ 1 \end{array}$$

That's it

Mental Calculation

Let's calculate $72 \div 4$ mentally.

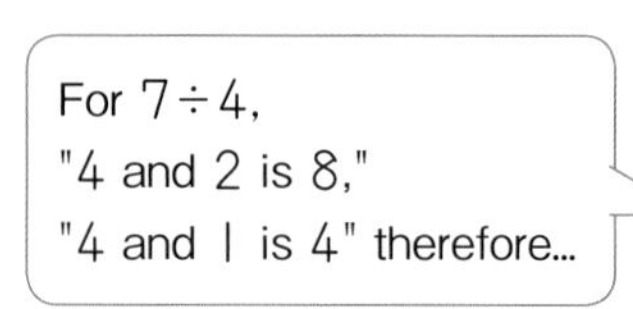

$72 \div 4$ <
- $40 \div 4$ → "4 and 1 is 4" → 10
- $32 \div 4$ → "4 and 8 is 32" → 8

> → Altogether ☐

5 The case: (3-digit) ÷ (1-digit) = (2-digit)

Want to solve

1 **We want to divide 254 sheets of colored paper such that 3 children receive the same number. How many sheets of paper will each child receive and how many sheets of paper will remain?**

① Let's write a math expression.

② Let's think about the division algorithm in vertical form.

Purpose What should we do in a calculation when the quotient cannot be written in the hundreds place?

Division algorithm to calculate 254 ÷ 3 in vertical form

3) 2 → 3) 2 5 → 3) 2 5 4 → 3) 2 5 4

$$\begin{array}{r} 84 \\ 3\overline{)254} \\ 24 \\ \hline 14 \\ 12 \\ \hline 2 \end{array}$$

2 ÷ 3
We cannot write a quotient in the hundreds place.

25 ÷ 3
Write a quotient in the tens place.

Summary

When the quotient is not written in the hundreds place, the calculation starts from the tens place.

③ Let's confirm the answer.

Want to confirm

Let's solve the following calculations in vertical form.

① 316 ÷ 4　② 552 ÷ 6　③ 329 ÷ 7　④ 624 ÷ 8

⑤ 173 ÷ 2　⑥ 581 ÷ 9　⑦ 236 ÷ 3　⑧ 488 ÷ 5

6 Deciding which operation

Let's find the answer of the following problem.

① If each group is handed out with 18 sheets of colored paper and 7 groups were handed out, then how many sheets of colored paper were needed?

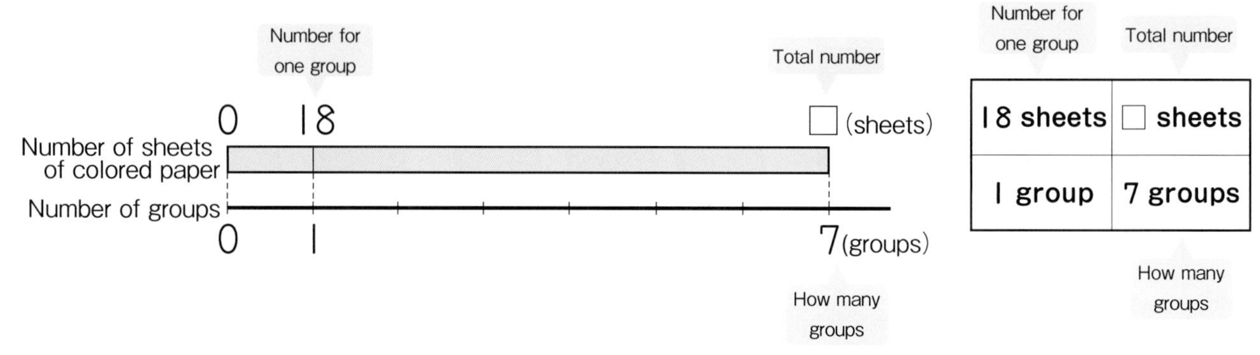

Number for one group / Total number

18 sheets	□ sheets
1 group	7 groups

How many groups

② If the total number of sheets is 126 and each group is handed out with 18 sheets, then how many groups will receive?

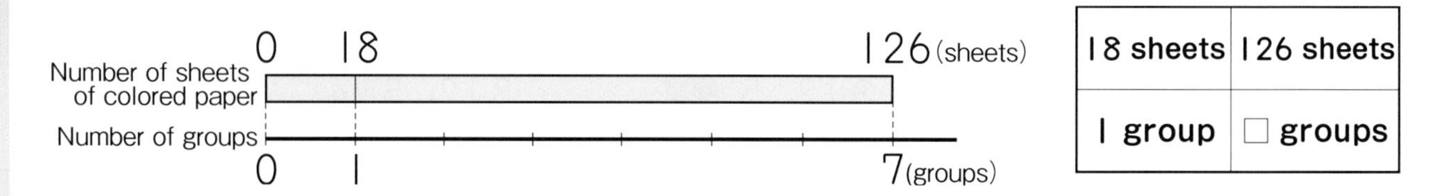

18 sheets	126 sheets
1 group	□ groups

③ If the total number of sheets is 126 and 7 groups were handed out with equal number of sheets, then how many sheets will each group receive?

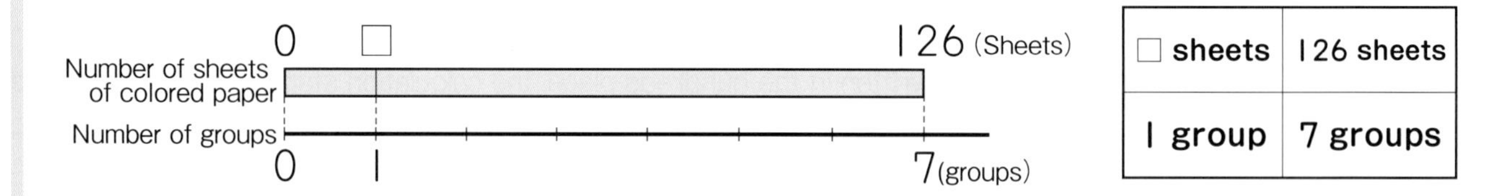

□ sheets	126 sheets
1 group	7 groups

Use multiplication to find the total number. Use division to find the number for one group and how many groups.

What you can do now

☐ **Understanding the division algorithm in vertical form.**

1 Let's think about the division algorithm to calculate $294 \div 3$ in vertical form.

① The quotient is written from the ☐ place value.

② The remainder 2 in the tens place represents 2 sets of ☐.

③ The calculation in the ones place is ☐ $\div 3$.

$3\overline{)2\ 9\ 4}$

☐ **Can solve 2-digit ÷ 1-digit divisions.**

2 Let's solve the following calculations in vertical form.

① $78 \div 3$	② $96 \div 8$	③ $38 \div 2$
④ $55 \div 5$	⑤ $48 \div 4$	⑥ $77 \div 6$
⑦ $56 \div 3$	⑧ $90 \div 7$	⑨ $83 \div 2$
⑩ $65 \div 3$	⑪ $98 \div 9$	⑫ $81 \div 4$

☐ **Can solve 3-digit ÷ 1-digit divisions.**

3 Let's solve the following calculations in vertical form.

① $548 \div 4$	② $259 \div 7$	③ $624 \div 3$
④ $367 \div 9$	⑤ $457 \div 6$	⑥ $543 \div 5$
⑦ $963 \div 8$	⑧ $728 \div 6$	

☐ **Can make a division expression and find the answer.**

4 Mitsuki and her 5 friends will fold 360 cranes. If each one folds the same number of cranes, how many cranes should each one fold?

Supplementary Problems p.151

Usefulness and efficiency of learning

1 Let's solve the following calculations in vertical form.

① $34 \div 4$　② $72 \div 5$　③ $86 \div 2$

④ $70 \div 5$　⑤ $174 \div 6$　⑥ $589 \div 7$

⑦ $828 \div 3$　⑧ $914 \div 7$　⑨ $750 \div 8$

- Understanding the division algorithm in vertical form.
- Can solve 2-digit ÷ 1-digit divisions.
- Can solve 3-digit ÷ 1-digit divisions.

2 There are 125 children that must race in groups of 6 children.

① How many groups of 6 children are there?

② If the remaining children form a group, how many children race in that group?

- Can make a division expression and find the answer.

3 There are 436 pencils. The pencils are devided into sets of 3 pencils.

How many sets of 3 pencils can be made?

Also, how many pencils are needed to make 150 sets?

- Can make a division expression and find the answer.

4 You must create a square using a 64 cm string.

How many cm is the length of one side?

- Can make a division expression and find the answer.

5 Let's find all the whole numbers that divided by 6 result with a quotient 8.

Let's deepen.

As for the division, in what type of daily situation can be used?

Deepen. How many trees?

Want to know

In Manganji (Niigata City, Niigata Prefecture), Hasaki is a tree that is planted along the rice fields in order to dry rice after mowing.

Let's consider the case in which these trees are planted in a line like this.

Hasaki in Manganji (Niigata City, Niigata Prefecture)

Want to think

① In a road's length of 80 m, as shown on the right, the trees are planted every 4 m. How many trees are there?

② In a road's length of 1000 m, trees are planted every 2 m. How many trees are there?

Do I also have to think about the trees at both ends?

③ On exercise ②, if trees are planted on both sides of the road, what will be the total number of trees?

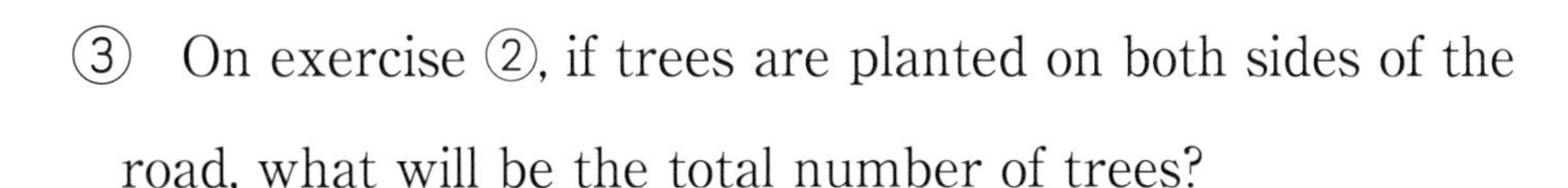

5 Angles
Let's think about how to measure and draw angles.

1 Size of an angles

Want to exploe Animal's way to open the mouth

Activity

The animals from Ⓐ~Ⓔ have their mouths open. Let's explore how the animals open their mouth.

① Which animal opens the mouth the widest?

② Which animal opens the mouth the narrowest?

The amount of space formed by rotating from one side to another side is called **size of an angle**.

size of an angle

③ Let's name the angles in order of the size of the angle.

Want to think

④ Let's think about how to compare the size of an angle.

Daiki's idea

I trace an angle on a clear paper and compare them by placing one over the other.

Nanami's idea

I examine by how many times is the size of an angle in the triangle ruler.

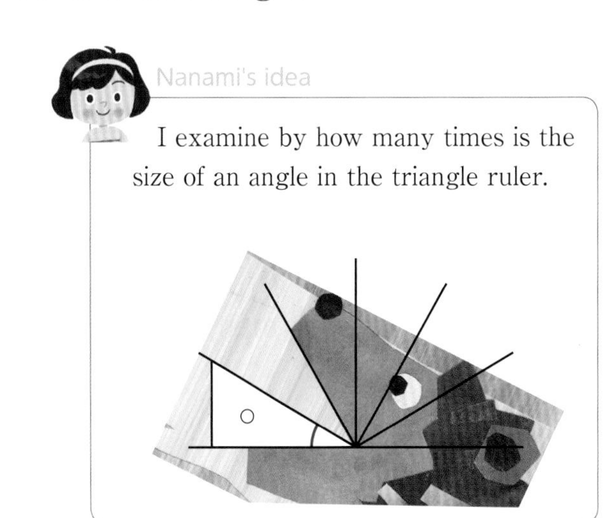

The size of an angle is not related to the length of the sides, it is determined by the opening between the sides.

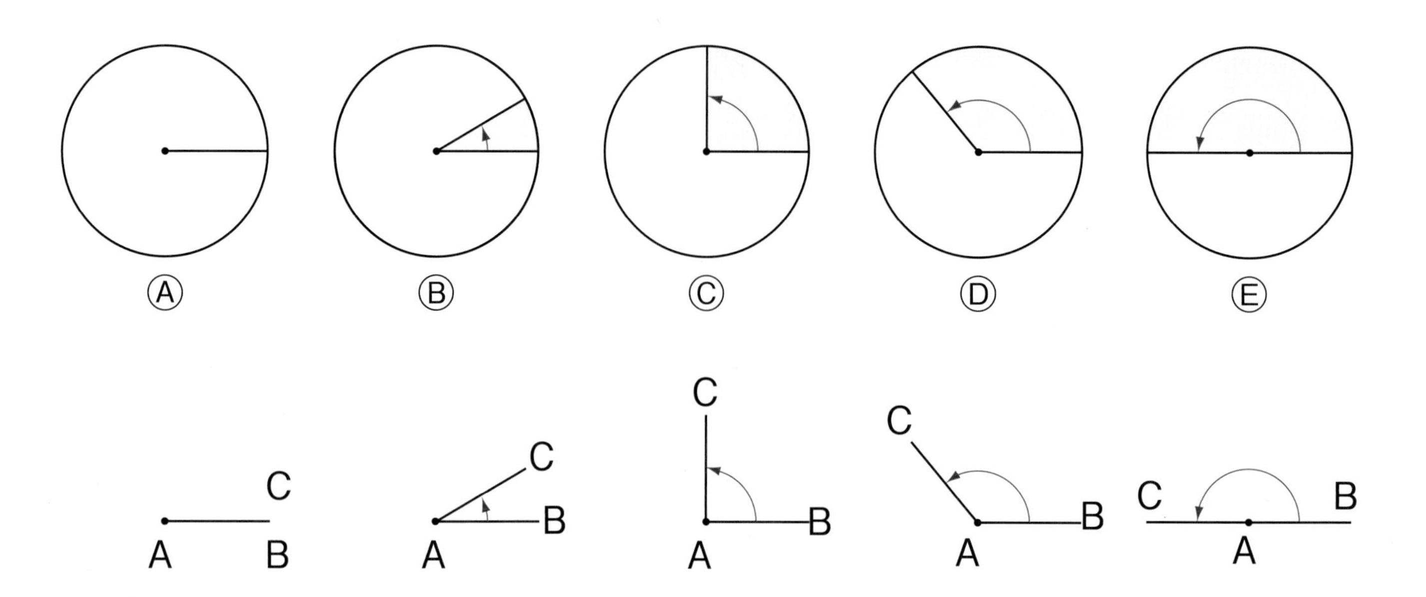

2 The size of rotating angles

Want to explore

1 **Let's explore the rotated angle shown on the above diagram.**

① Centered on point A, if side AC rotates on the direction of the arrow, how does the size of the angle change?

② From the angles Ⓐ~Ⓘ, which became a right angle?

③ How many right angle parts make the size of angles Ⓔ, Ⓖ, and Ⓘ ?

The size of angle Ⓔ is made by 2 right angle parts, so is called 2 right angles. Also, angles Ⓒ, Ⓖ, and Ⓘ are called, respectively, 1 right angle, 3 right angles, and 4 right angles. Additionally, 4 right angles are called "angle of one revolution" and 2 right angles are called "angle of a half revolution."

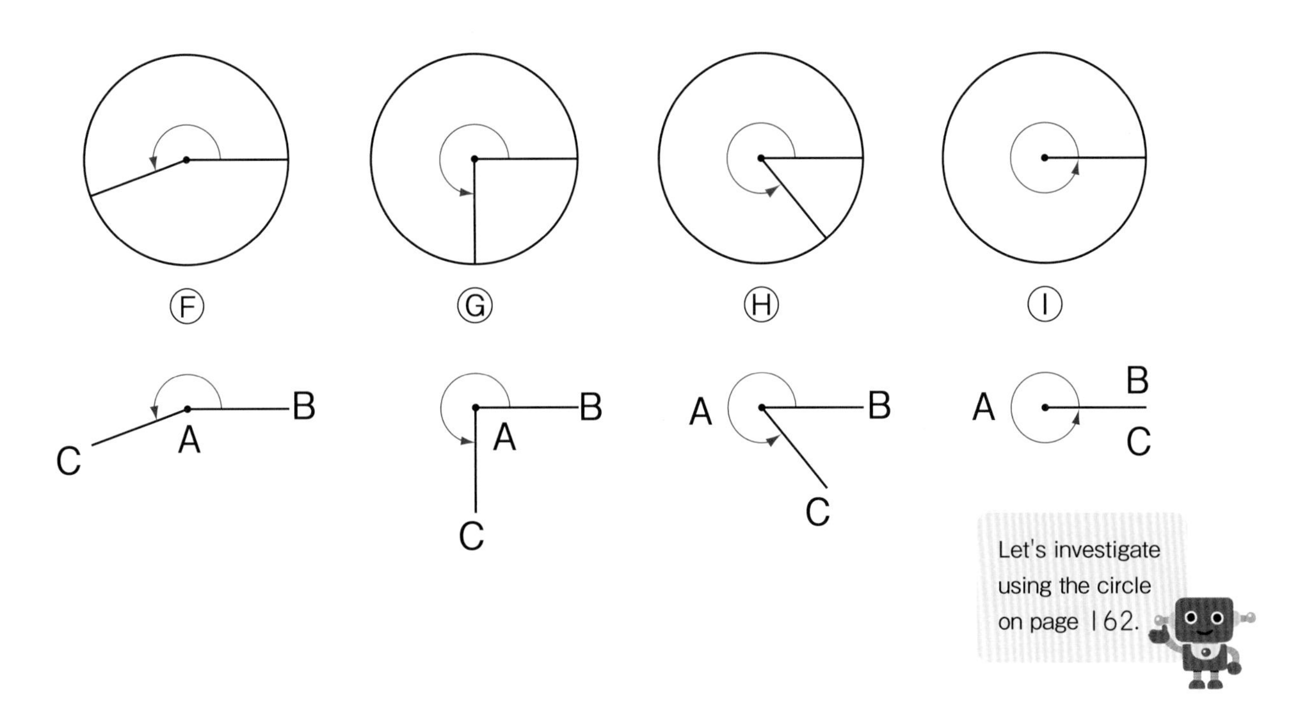

Let's investigate using the circle on page 162.

Want to improve

④ Let's use a triangle ruler to make the size of angles Ⓑ and Ⓓ.

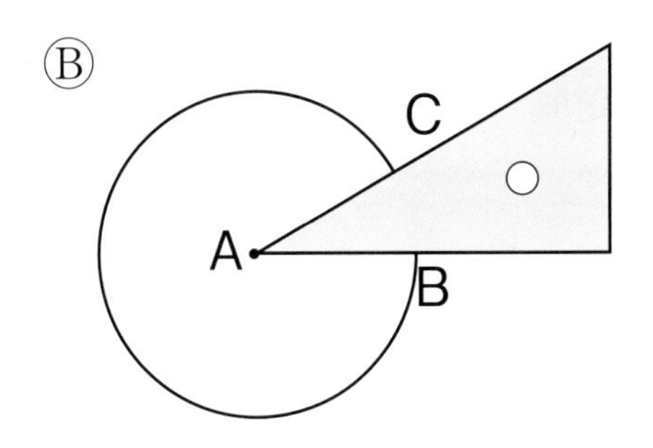

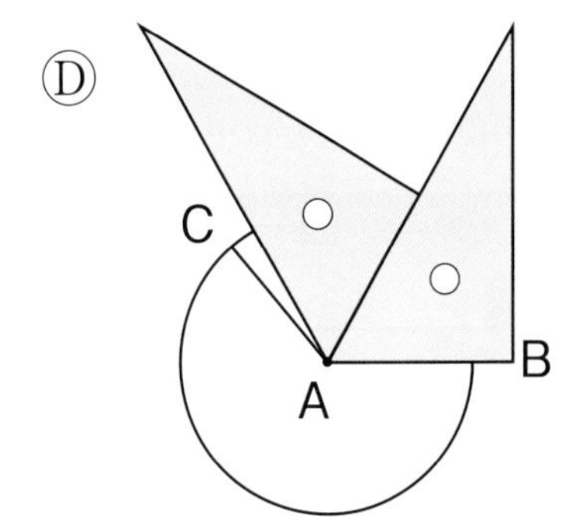

Yui: You can create angle Ⓑ by using one angle from a triangle ruler.

Hiroto: You can not exactly create angle Ⓓ by using triangle rulers.

Purpose Can we represent the size of an angle with a number?

❸ How to measure angles

Want to know How to use a protractor

A protractor is used to correctly measure the size of angles.
Let's explore about how is the scale of a protractor.

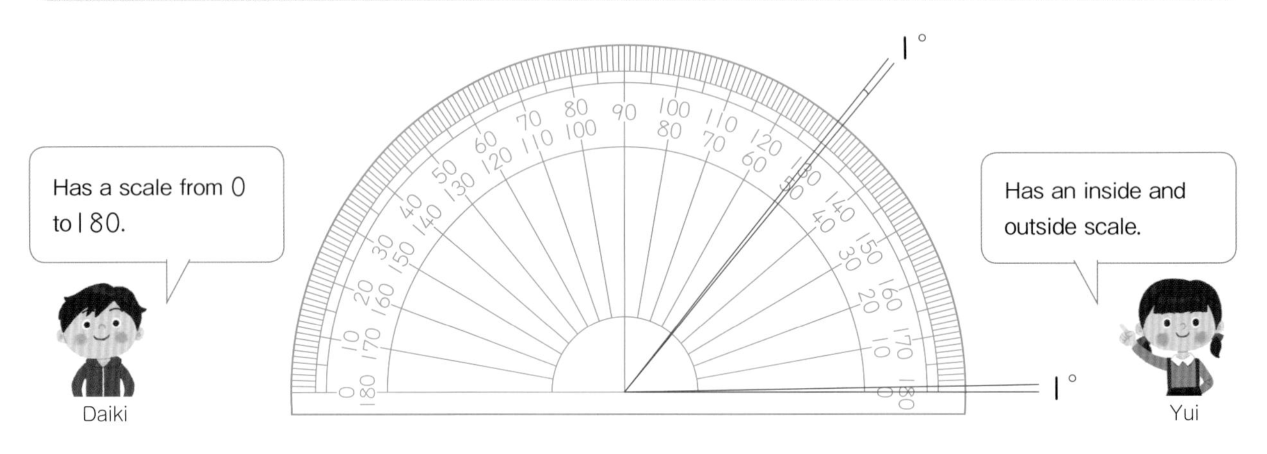

One angle revolution has 360 equal parts. The size of an angle with one of those equal parts is written as 1° and is read as 1 **degree**. Degree is the unit for the size of an angle. The size of an angle is called **angle**.

Summary

As for the size of an angle, whatever part can be represented by using the 1° unit.

① How many degrees are the angles Ⓔ, Ⓖ, and Ⓘ in p. 60 - 61?

1 right angle = 90°

Ⓔ 2 right angles = ☐°

Ⓖ 3 right angles = ☐°

Ⓘ 4 right angles = ☐°

Ⓔ 

Ⓖ 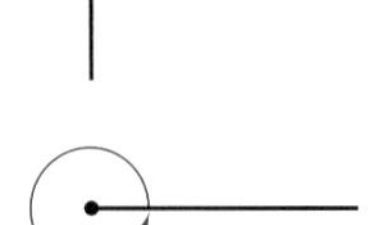

 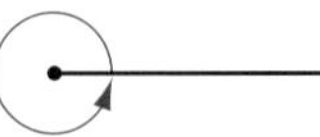

Using a protractor, let's measure angle ⓐ shown on the right.

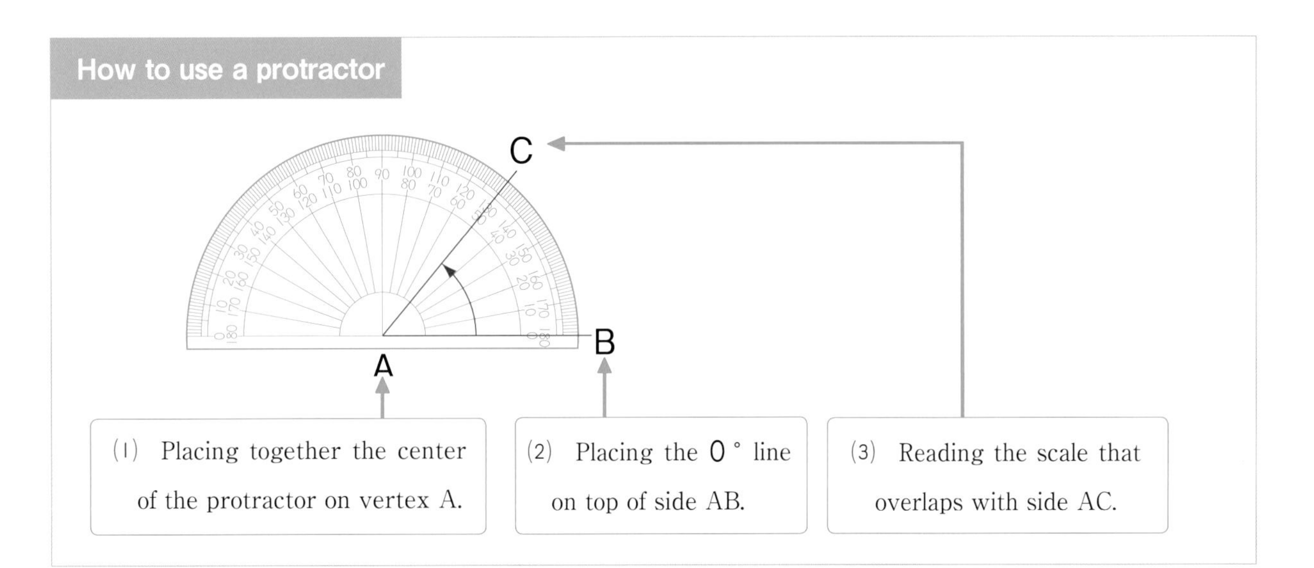

The angle ⓐ is 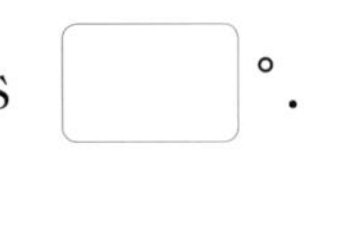 °.

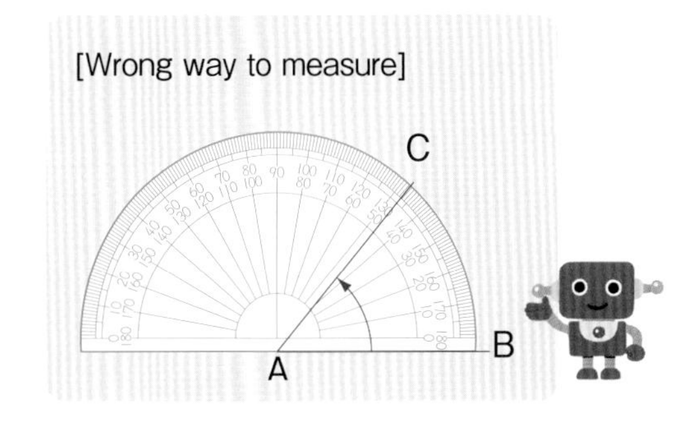

Want to confirm

How many degrees are the following angles?

①

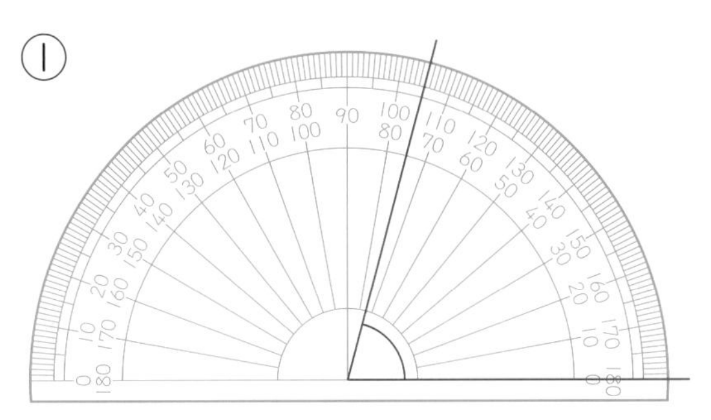

②

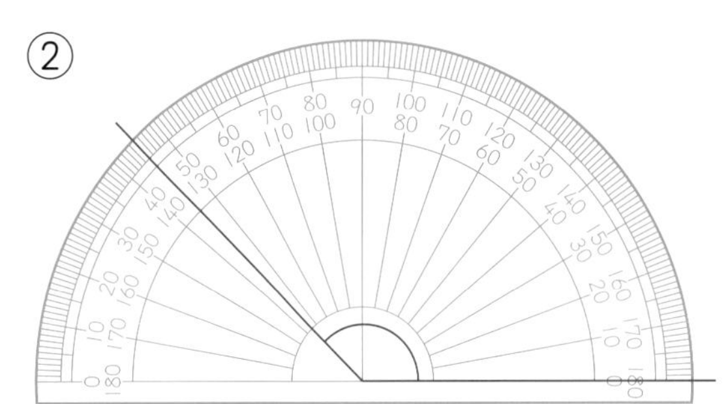

Want to think

2 Let's think about how to measure the following angles.

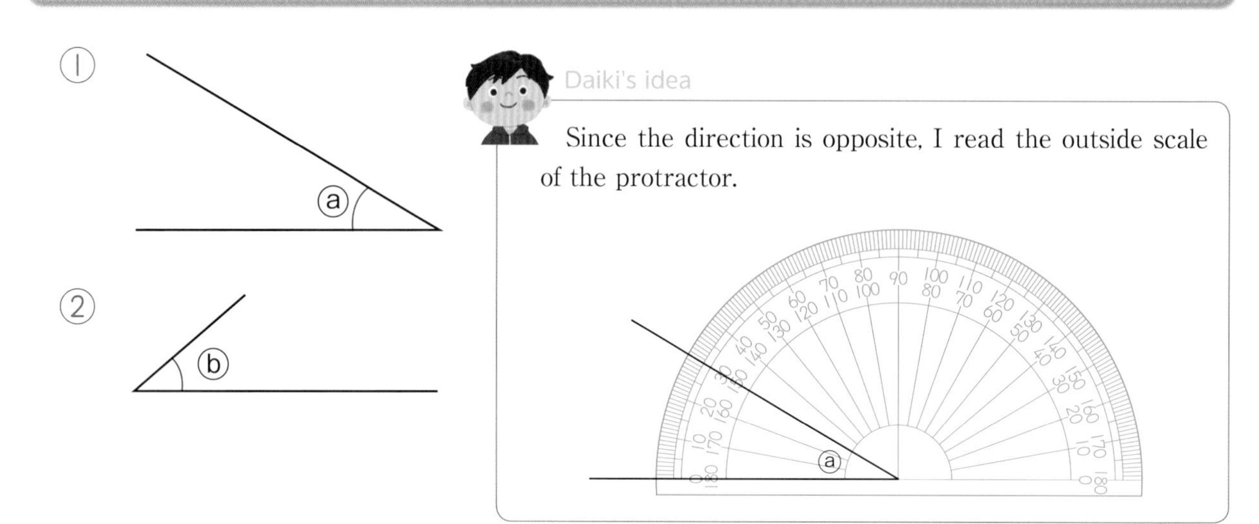

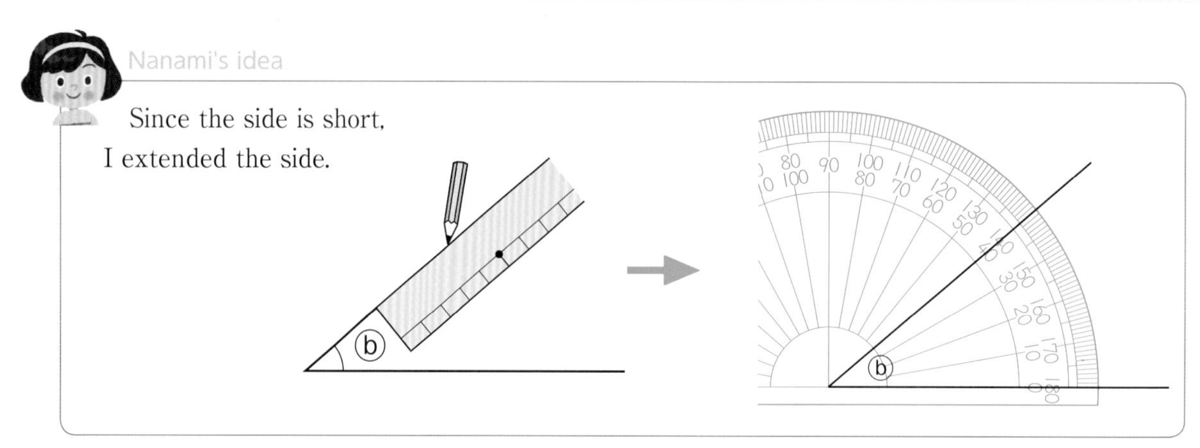

Want to confirm

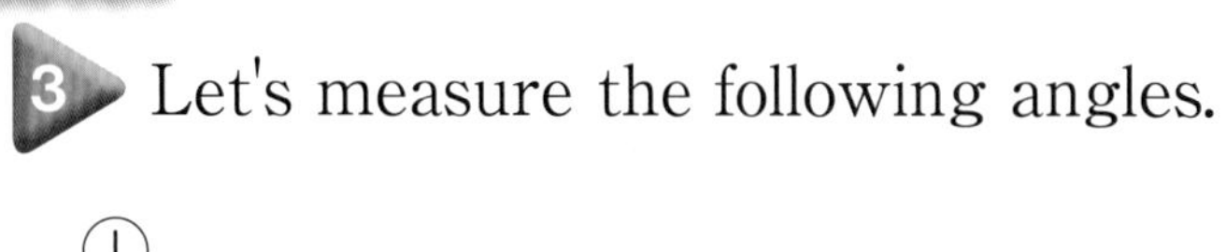

3 Let's measure the following angles.

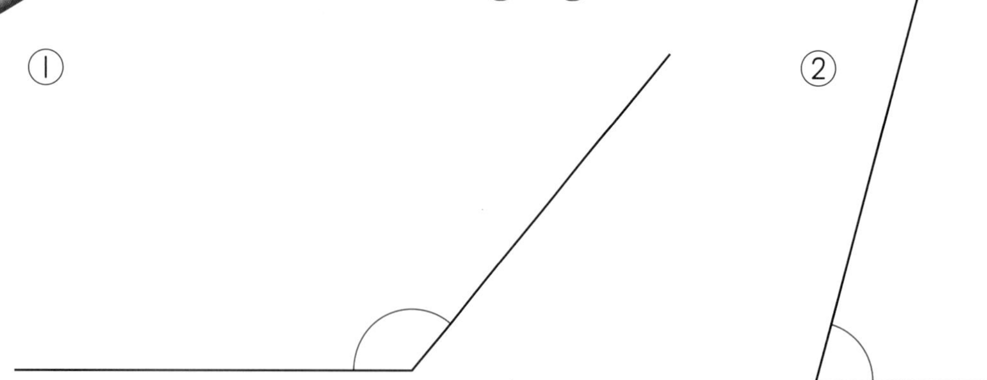

Want to try

4 Let's measure the following angles.

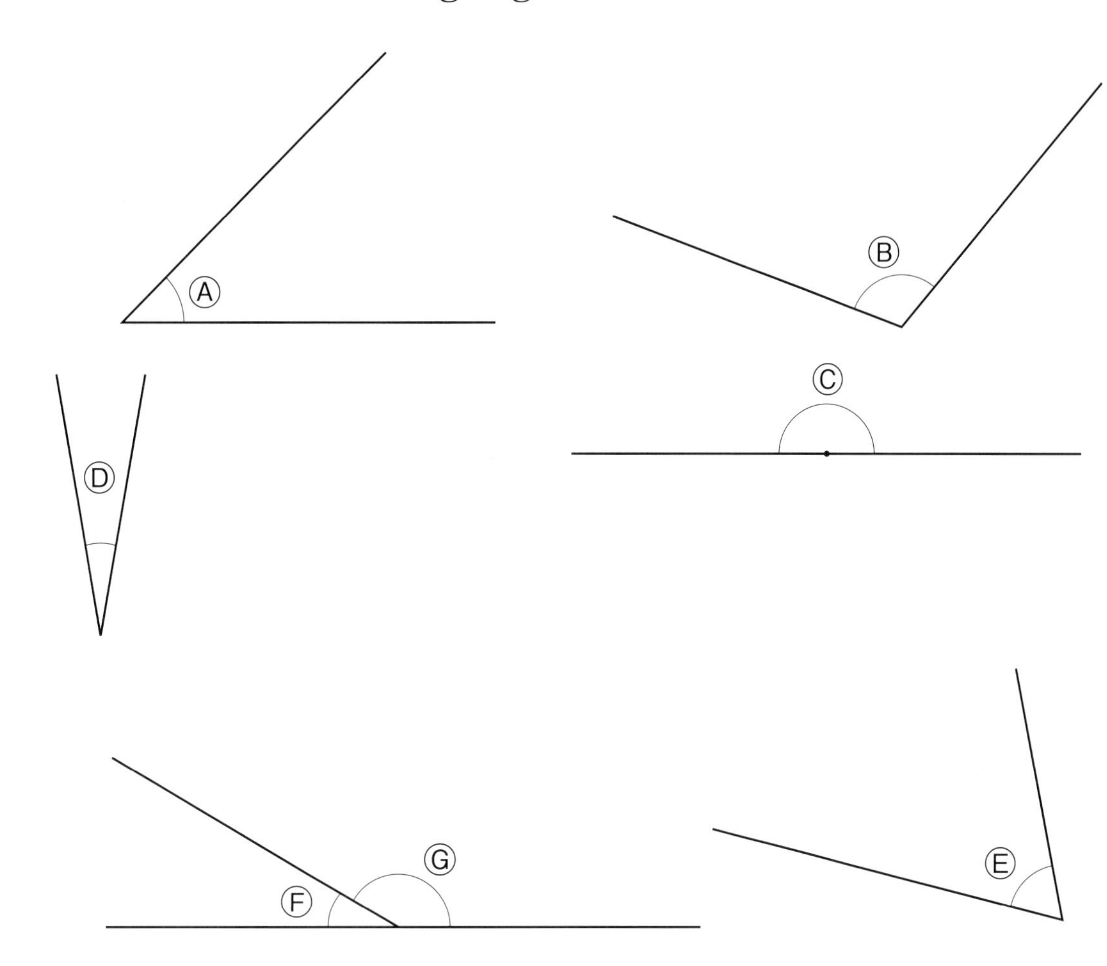

Want to improve

3 Let's measure angle ⓐ with a new idea.

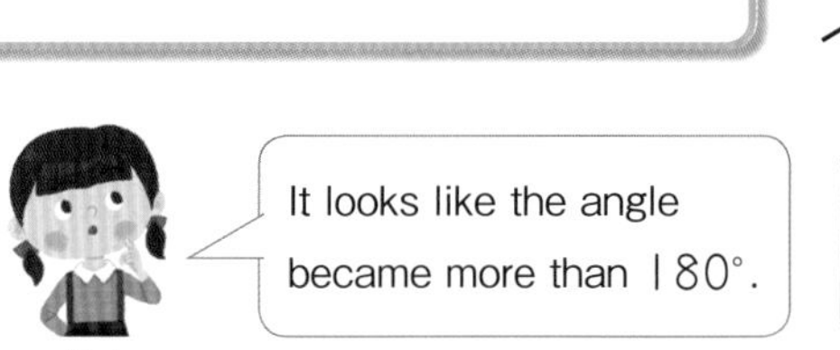

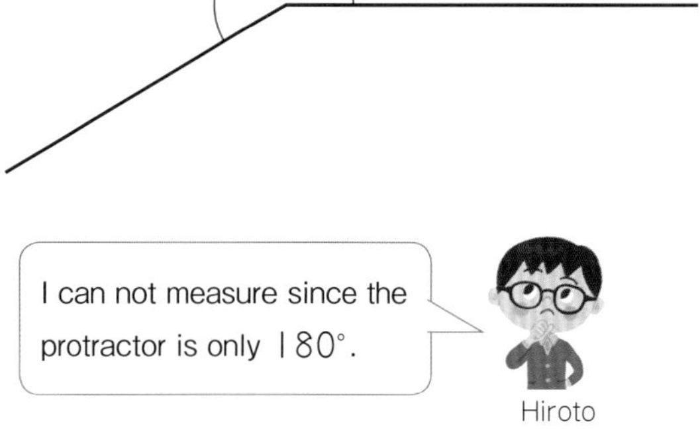

Purpose How do we measure an angle larger than 180°?

Want to discuss

① Let's discuss the new ideas from Yui and Hiroto.

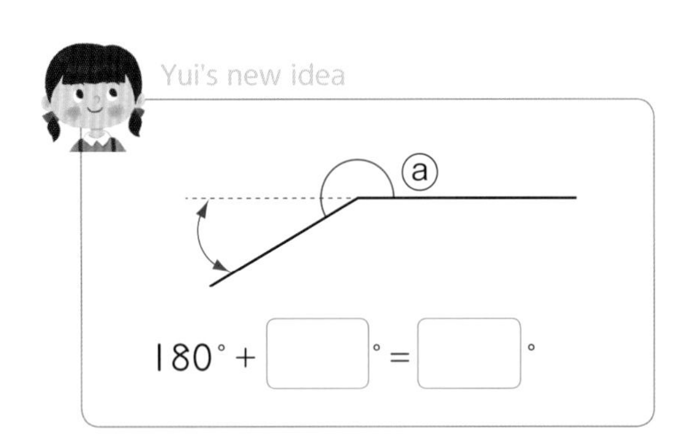

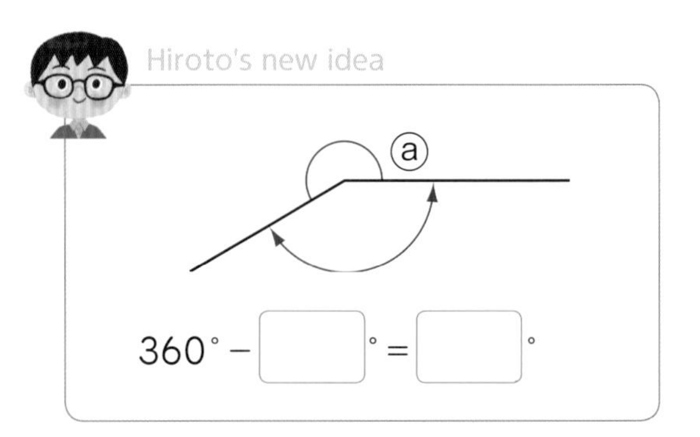

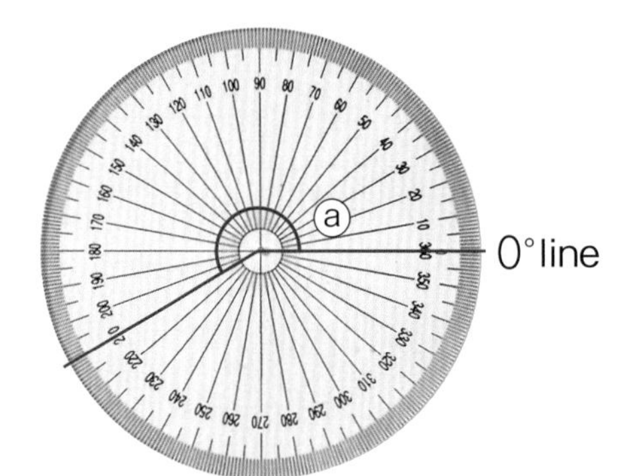

If you use an all round protractor, then you can measure it at once.

Summary

When you measure an angle larger than 180°, think how many degrees larger it is than 180°, or how many degrees smaller it is than 360°.

Want to confirm

5 How many degrees is the angle shown on the right?

Want to try

6 The diagram on the right shows the angle formation by the intersection of 2 straight lines. Let's explore about it.

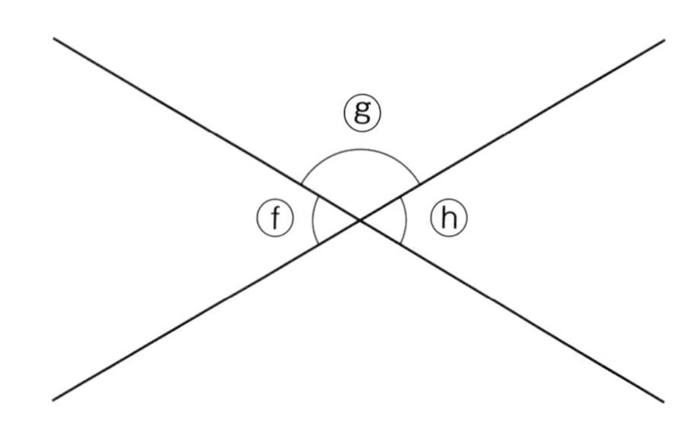

① If angle ⓕ is 60°, then, how many degrees is angle ⓖ?

② Let's compare angles ⓕ and ⓗ.

4 How to draw angles

Want to represent

1 **Using a protractor, let's draw an angle with size 50°.**

How to draw angles

(1) Drawing one side AB.

(1)
A B

(2) Placing together the center of the protractor and point A. Aligning the 0° line with side AB.

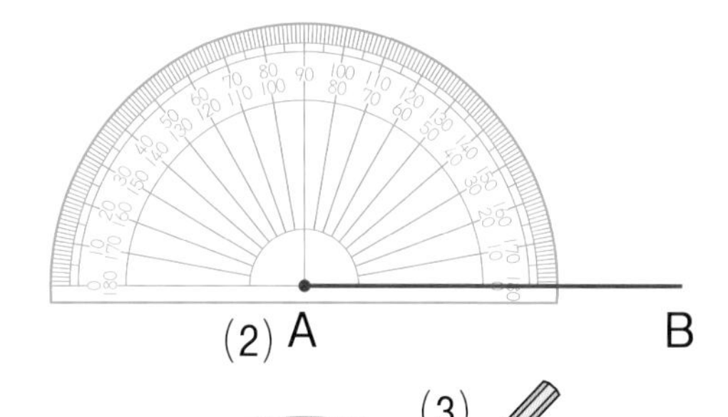

(3) Placing point C in the 50° scale place.

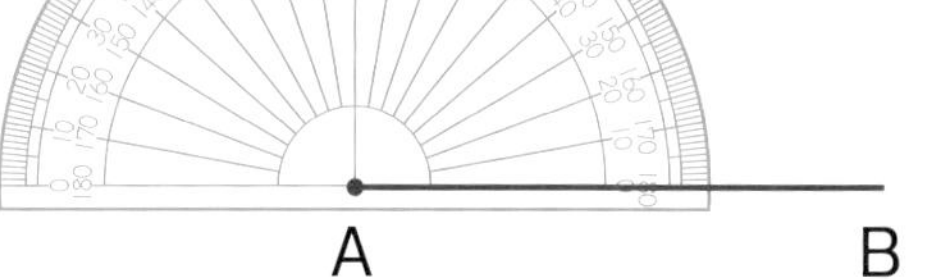

(4) Drawing a straight line through points A and C.

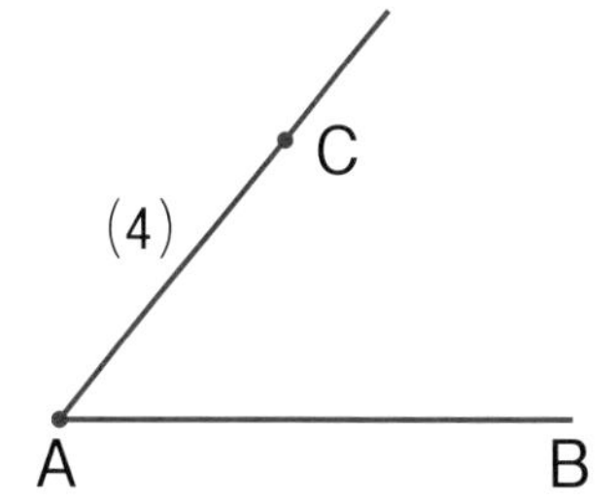

Want to confirm

Let's draw angles with size 85° and 140°.

Want to try

Let's draw an angle with size 210° using learned ideas.

Can I use the way to measure angles larger than 180°?

Daiki

Want to think How to draw triangles

Let's draw the triangle shown on the right.

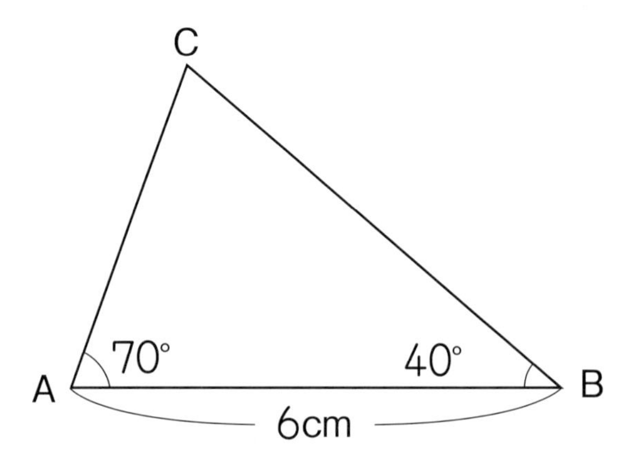

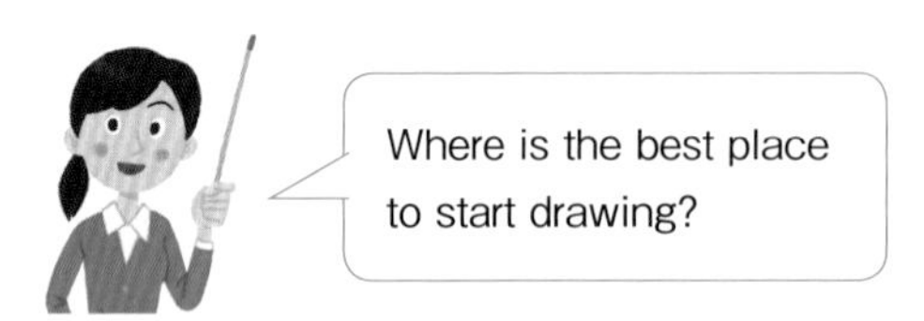

How to draw triangles

(1) Draw side AB with a length of 6 cm.

A ——— B

After this, better use the protractor.

(2) Place together the protractor center and point A. Draw a 70° angle.

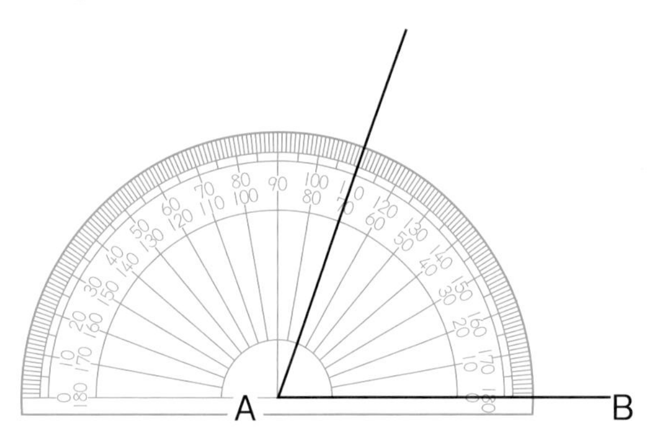

(3) Place together the protractor center and point B. Draw a 40° angle. Point C is the intersection point.

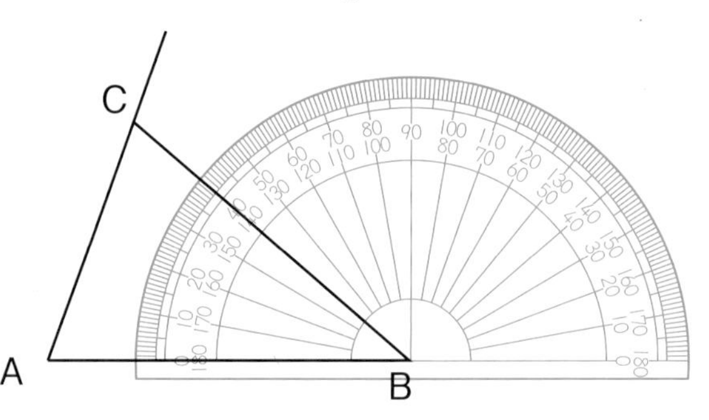

Want to confirm

Let's draw the following triangles.

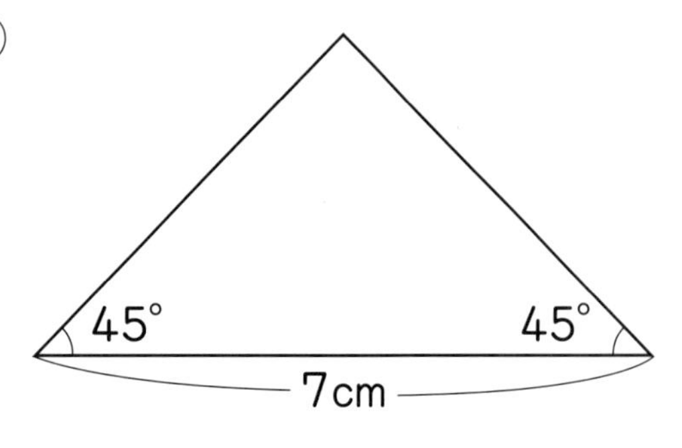

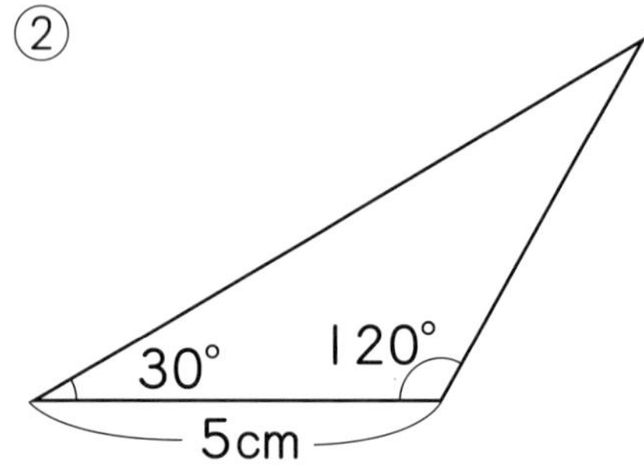

Making a pattern using a protractor

① Let's divide the circle in order of 10°, 20°, 30°, ..., until it overlaps the start line.

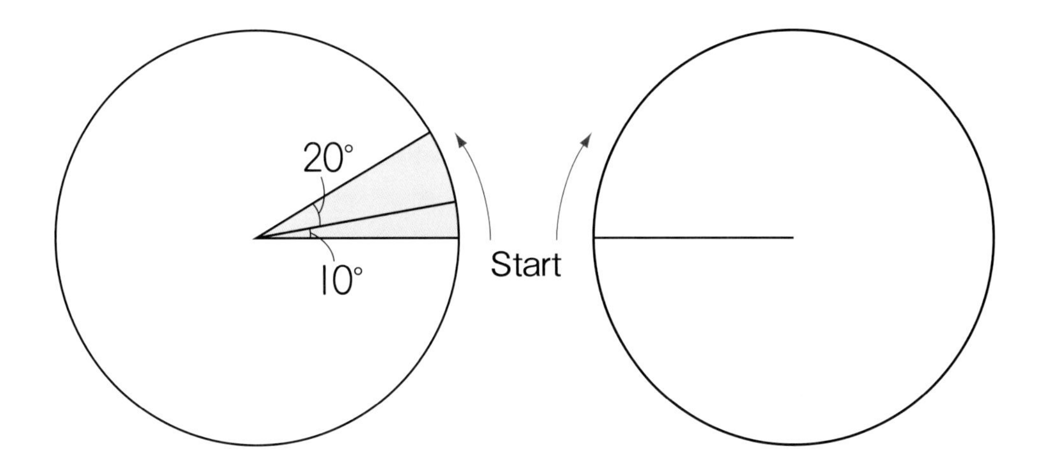

② From the starting point, go 6 cm to the right and turn 45° to the left. Then, go again 6 cm and turn 45° to the left. Let's do this repeatedly.

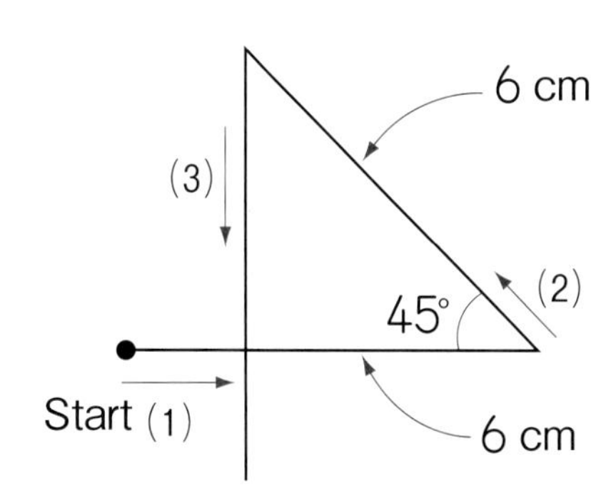

Start

5 Angles of triangle rulers

Want to know

Let's explore the size of angles in triangle rulers.

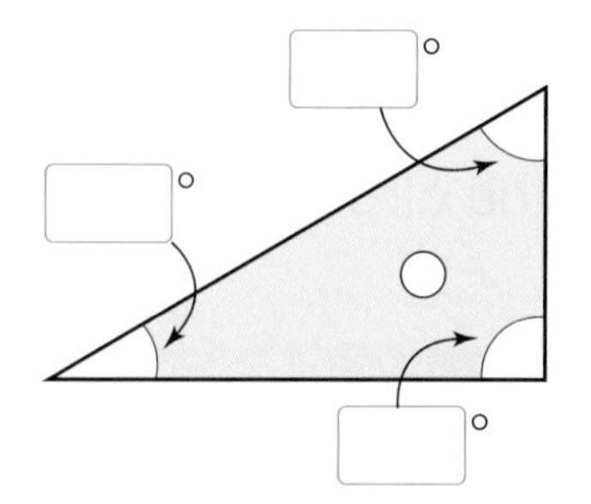

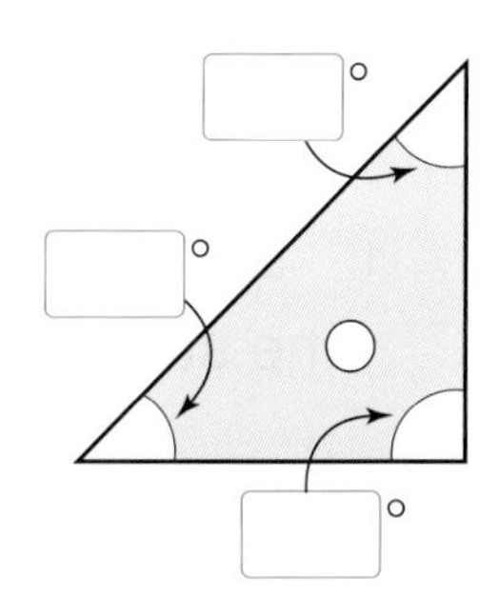

① Let's measure the size of each angle in one set of triangle rulers.

② Triangle rulers were combined to make angles. How many degrees is each angle?

Should the angles added or subtracted?

Daiki

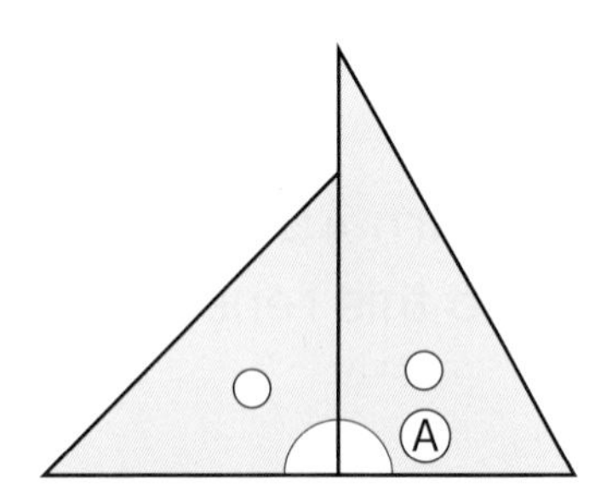

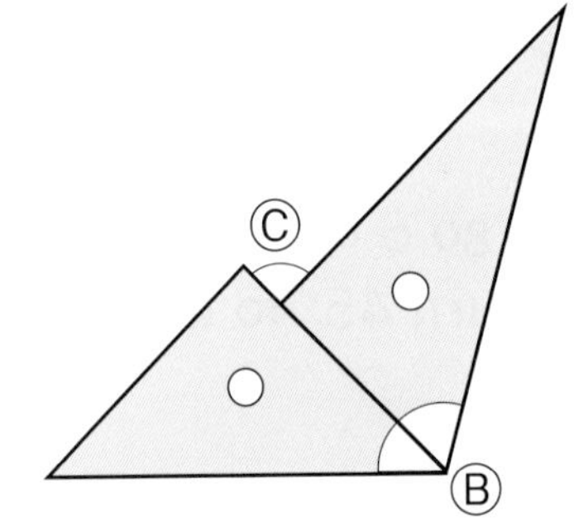

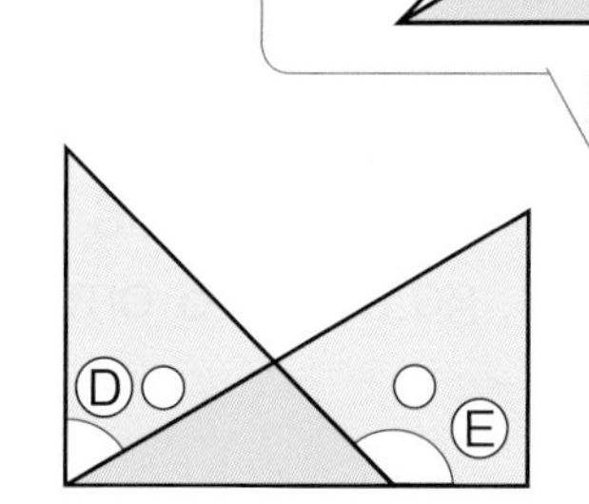

③ Let's combine triangle rulers to make a 150° angle.

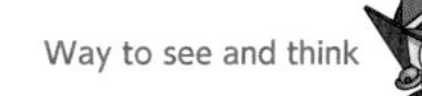

Way to see and think

The same way as for length and weight, you can also add and subtract the angle degree.

Want to find in our life

1 From your surroundings, let's investigate angles on various places.

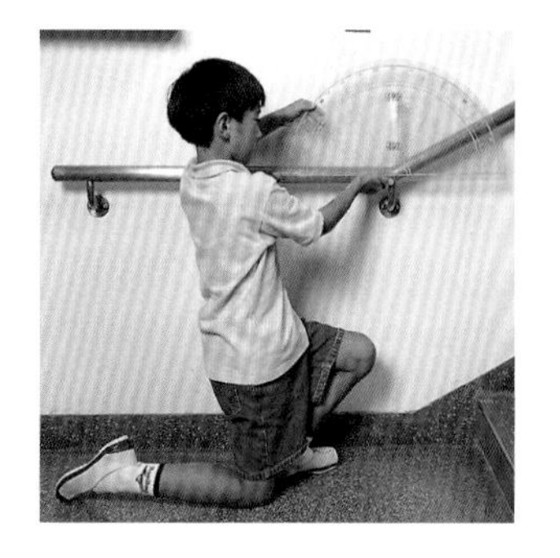

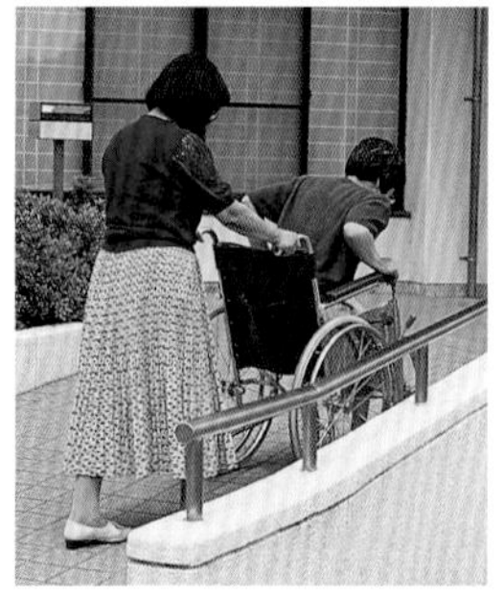

This shows that a track rises 10.5 m for every 1000 m.

What you can do now

☐ **Understanding how to represent the size of an angle.**

1 Let's summarize the size of an angle.

① The unit to represent the size of an angle is ☐.

② One revolution angle has ☐ equal parts. The size of an angle with one single part is 1 degree.

☐ **Can measure the size of an angle using the protractor.**

2 Let's measure the following angles.

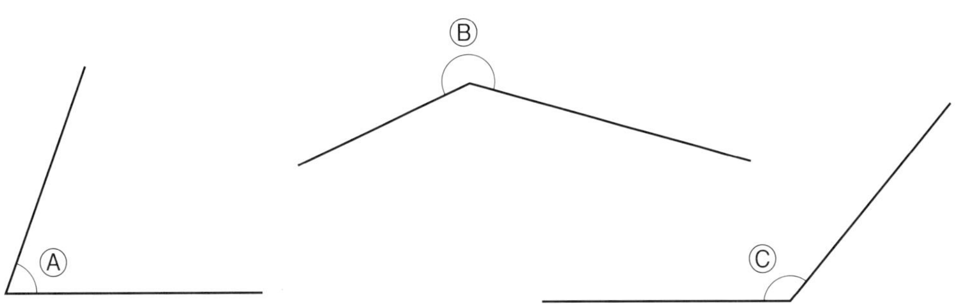

☐ **Can draw an angle using the protractor.**

3 Let's draw the angles with size 100° and 270°.

☐ **Understanding the angles of a triangle.**

4 Let's combine the triangle rulers to find the size of angles Ⓐ~Ⓒ.

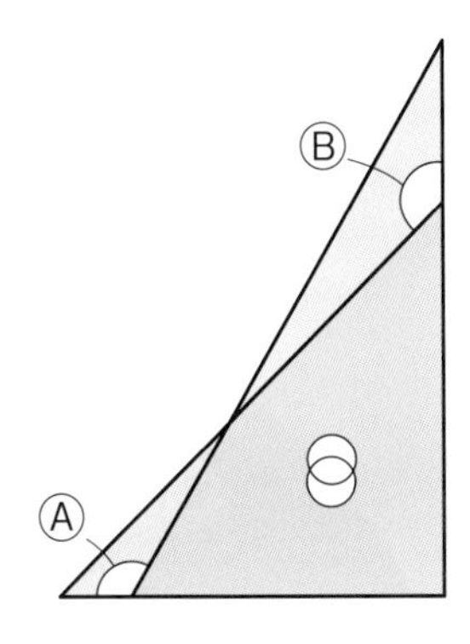

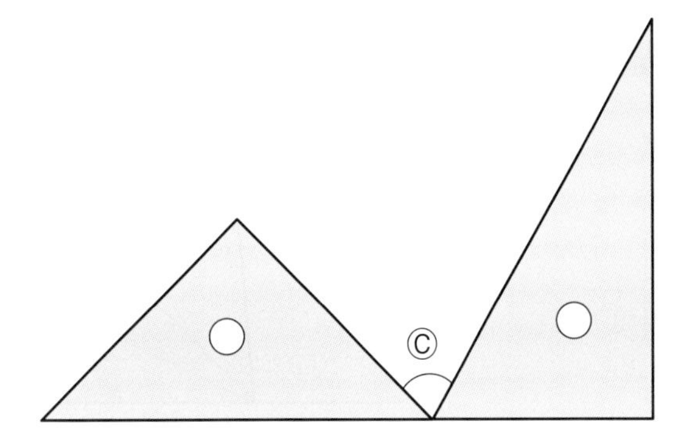

Supplementary problems ➡ p. 152

Usefulness and efficiency of learning

1 Let's measure the following angles.

☐ Understanding how to represent the size of an angle.

☐ Can measure the size of an angle using the protractor.

①

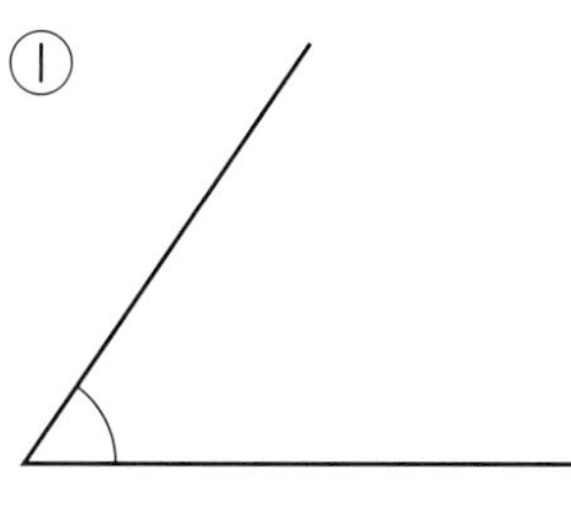

②

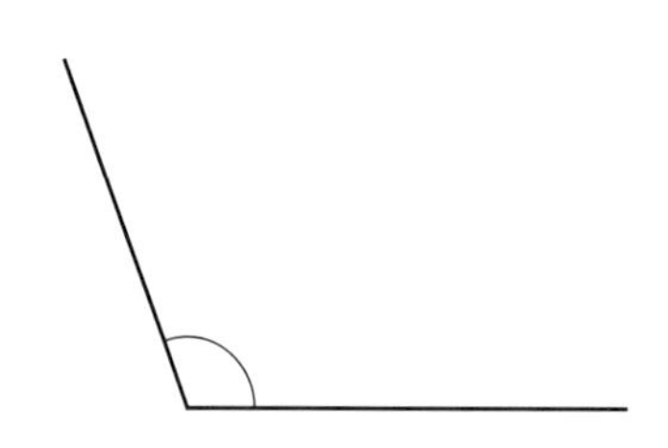

③

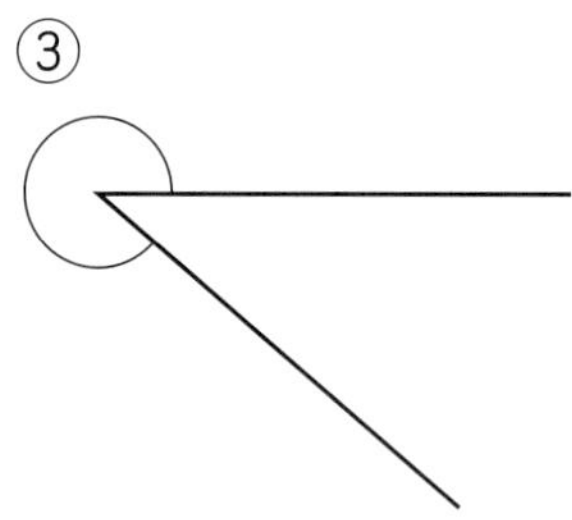

2 Let's draw angles with the following size.

☐ Can draw angles using the protractor.

① 120°　　② 300°

3 Let's combine the triangle rulers to find the size of angles Ⓐ~Ⓓ.

☐ Understanding the angles of a triangle.

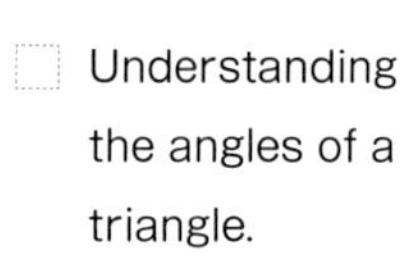

①

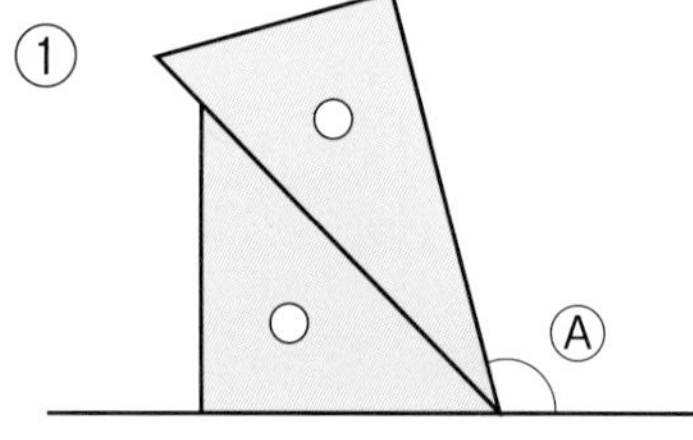

②

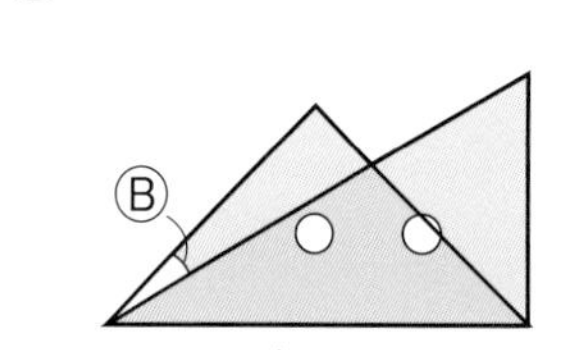

③

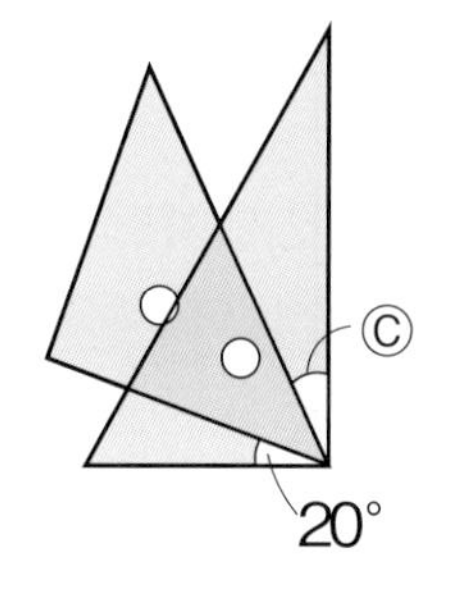

④

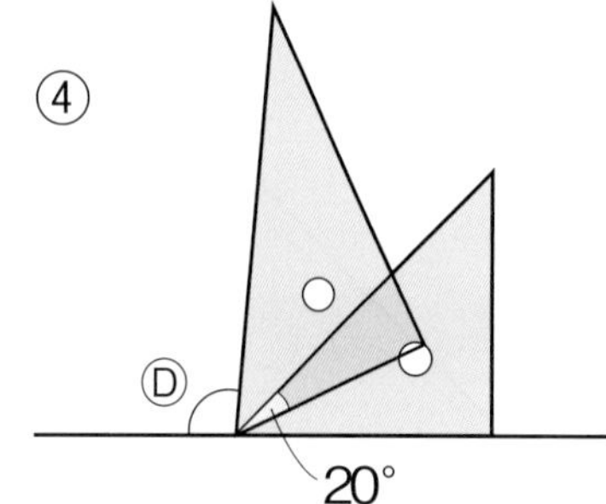

Let's deepen.

Is there a rule on the way clock hands open to make an angle?

Hiroto

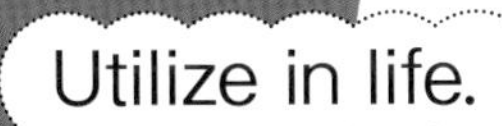

Deepen. Let's make a clock !

Want to know

Nanami thought about the angle formed at 4 p.m. between the long and short clock hand (shown on the right). Let's discuss her way of thinking.

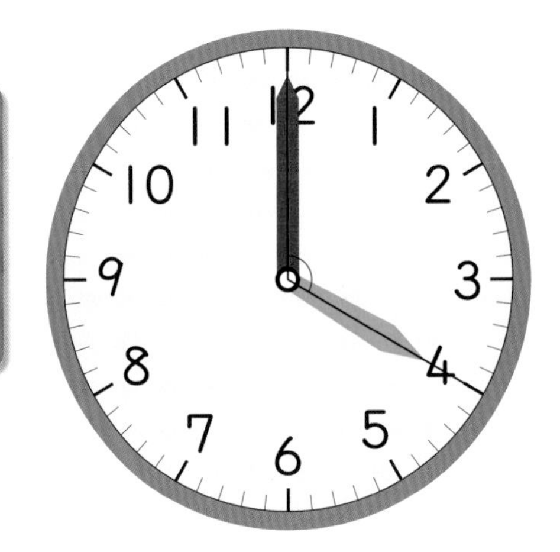

Nanami's idea

$180° \div 6 \times 4 = 120°$
therefore, became 120°.

180° is the size of the angle in the clock from the hour 12 to the hour 6.

As for 180° ÷ 6, what angle place is found?

Want to try

Using triangle rulers, you can create a 12 unit clock.
Let's write the numerals and make a clock.

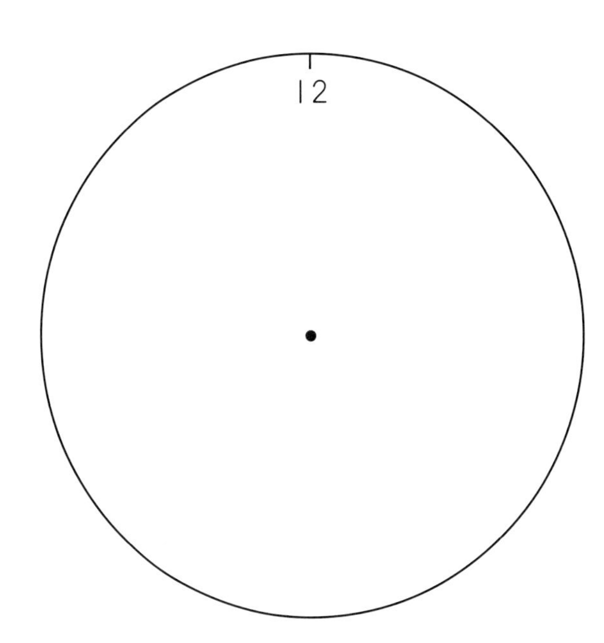

Daiki

If you think of the size of the angle between 12 and 1, you can make scales with triangle rulers.

What kind of figure is hidden?

Problem What kind of shapes are hidden in the straight line road?

Perpendicular, Parallel and Quadrilaterals

Let's explore the properties of quadrilaterals and classify them.

1 Perpendicular

Want to explore Creating right angles

By connecting points with 2 straight lines, let's make a right angle.

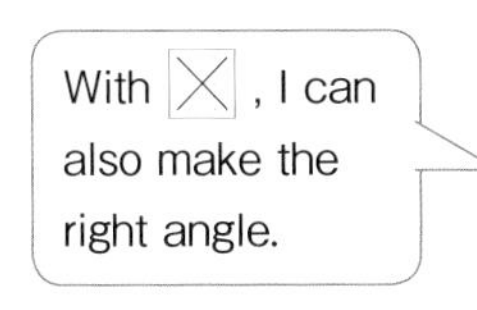

① Let's confirm the right angle was created with a triangle ruler or a protractor.

When the intersecting angle of 2 straight lines is a right angle, those 2 straight lines are **perpendicular** lines.

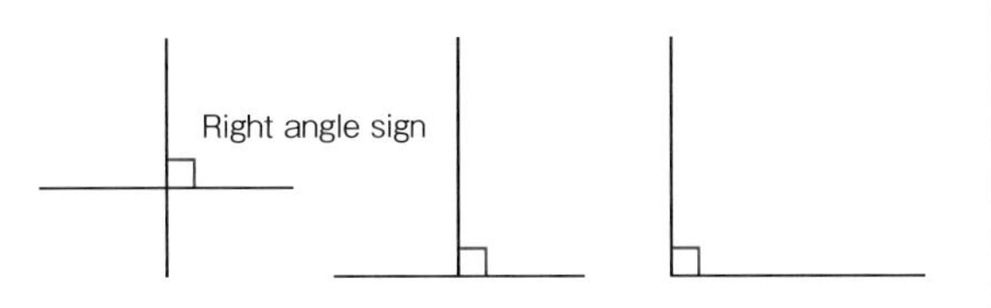

Words

 Hanging down

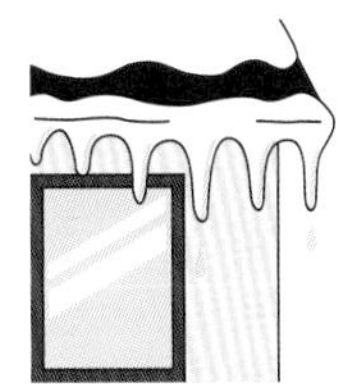

 Look straight

Want to explore Perpendicular lines when not intersecting

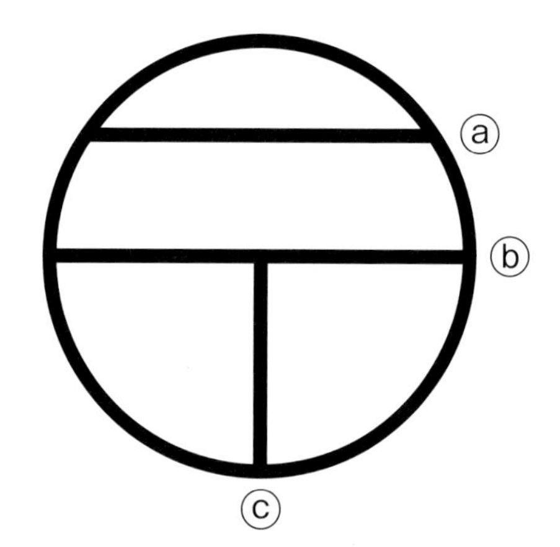

1 The diagram on the right shows the symbol for a post office in a Japanese map. Let's explore straight lines ⓐ, ⓑ, and ⓒ.

① At what angle do the straight lines ⓑ and ⓒ intersect?

② If you extend straight line ⓒ, how does it intersect with straight line ⓐ?

> Even if 2 straight lines do not intersect, when one or both straight lines are extended and intersect in a right angle, they are called perpendicular lines.

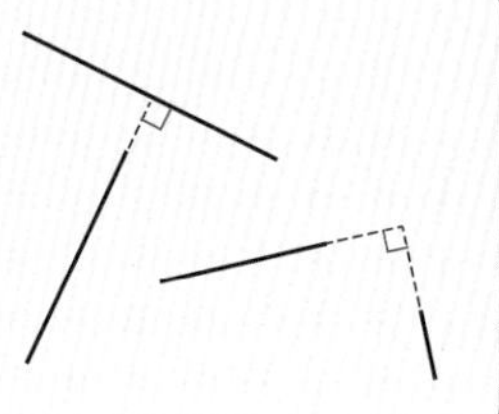

Want to confirm

2 Let's explore about perpendicular lines.

① In the following diagrams, which 2 straight lines are perpendicular lines?

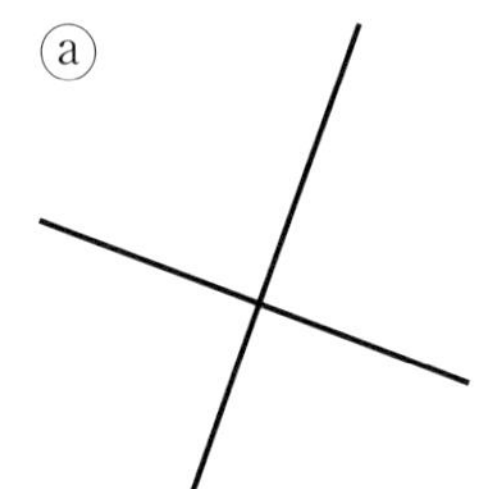

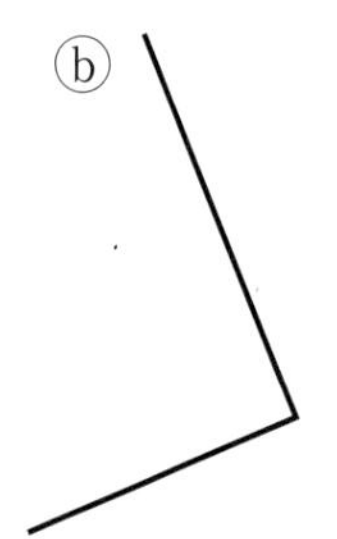

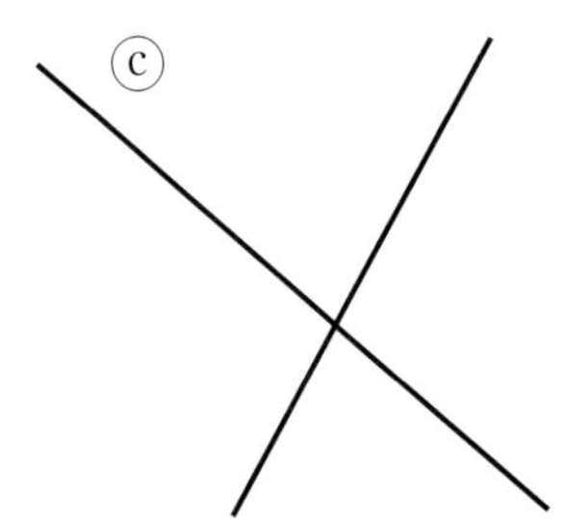

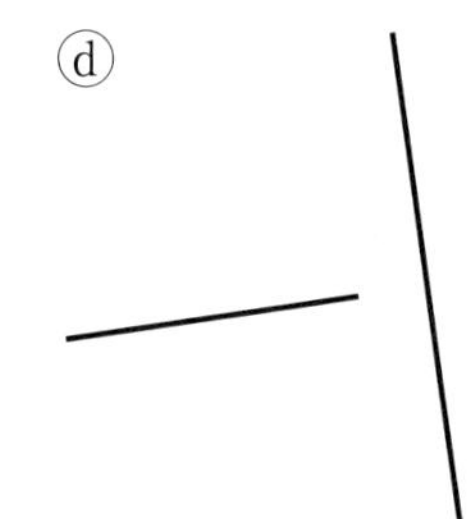

Want to try

② Let's fold a paper and make 2 straight lines that intersect as perpendicular lines.

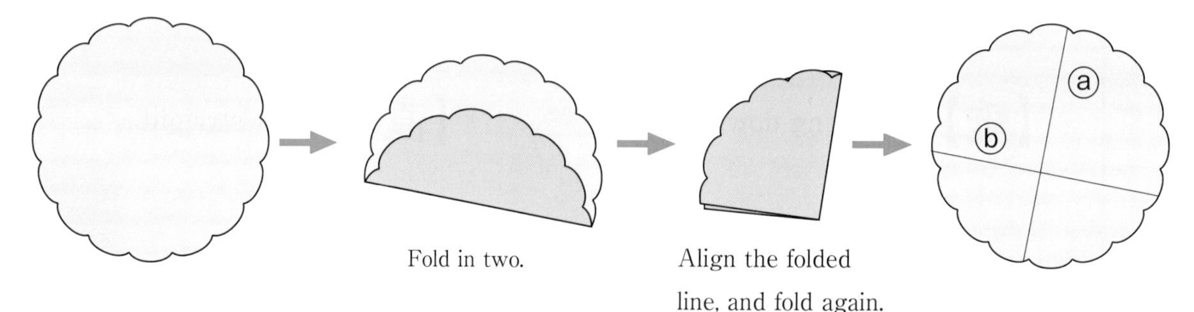

Fold in two.

Align the folded line, and fold again.

Want to explain How to draw perpendicular straight lines

Activity

2 **As shown in the figure below, Hiroto and Nanami drew a perpendicular straight line to another straight line. Let's explain the idea of both students.**

Hiroto's way of drawing

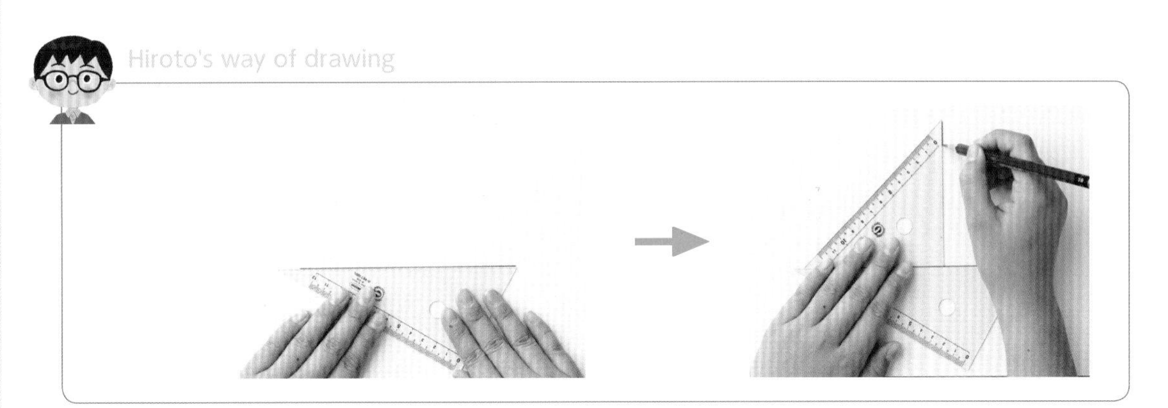

01401

Nanami's way of drawing

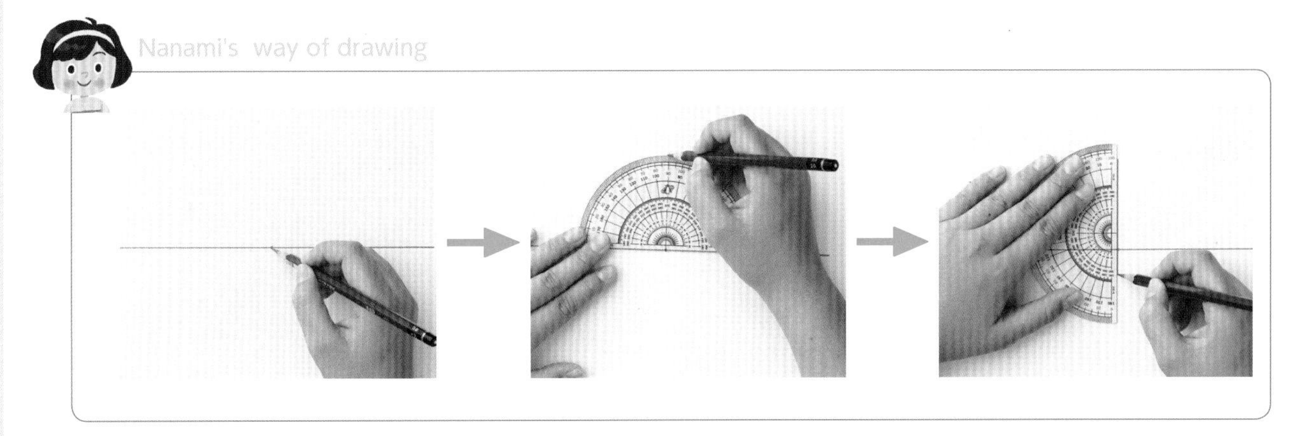

Want to try

3 Through point A, let's draw a perpendicular straight line to straight line ⓐ.

A •

ⓐ

How to draw perpendicular straight lines

When it goes through a point that is not on the straight line.

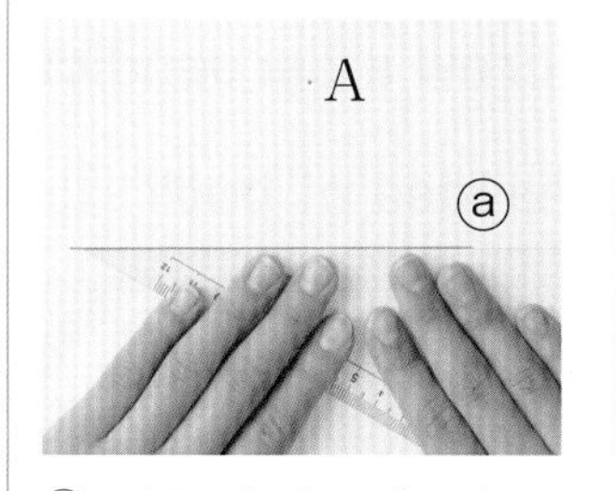

① Match the triangle ruler on straight line ⓐ.

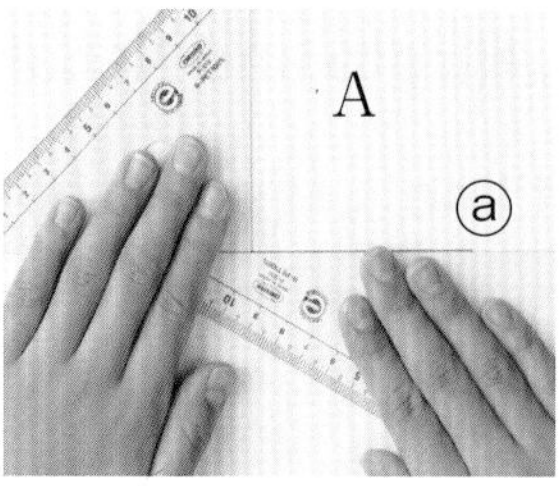

② Match the other triangle ruler's right angle side with straight line ⓐ.

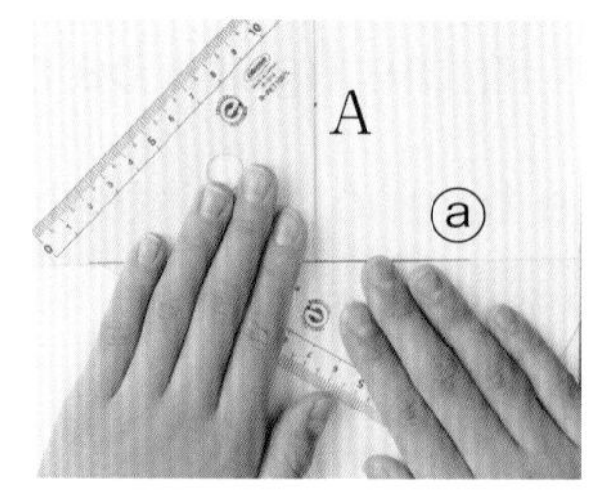

③ Move the above triangle ruler to match point A.

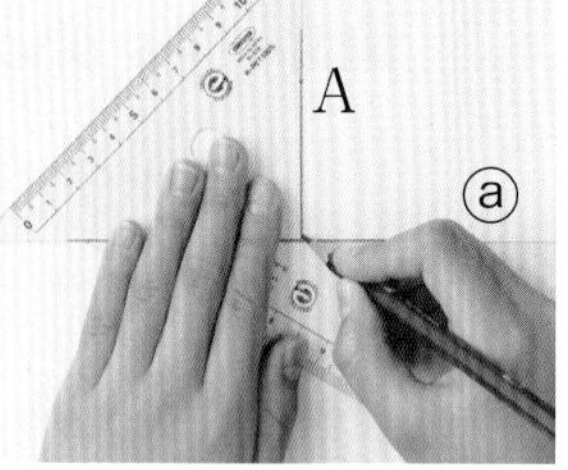

④ Holding the triangle ruler, draw a straight line passing through point A.

Want to try

4 Let's draw perpendicular straight lines to straight line ⓑ, through points B and C.

•B

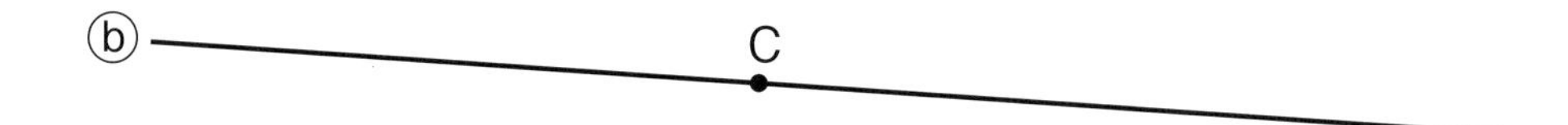

2 Parallel

Want to explore

Let's explore the arrangement of the following straight lines.

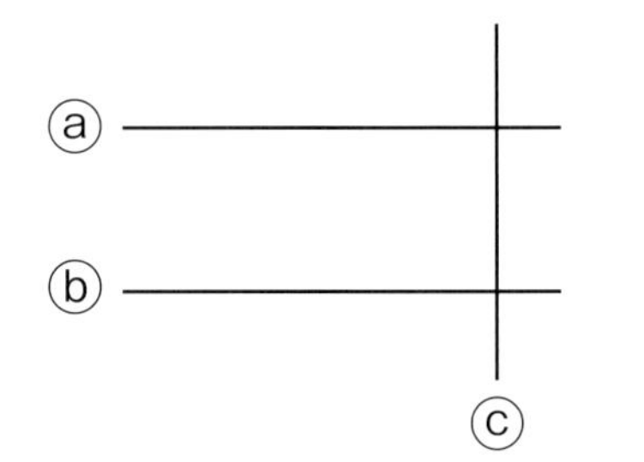

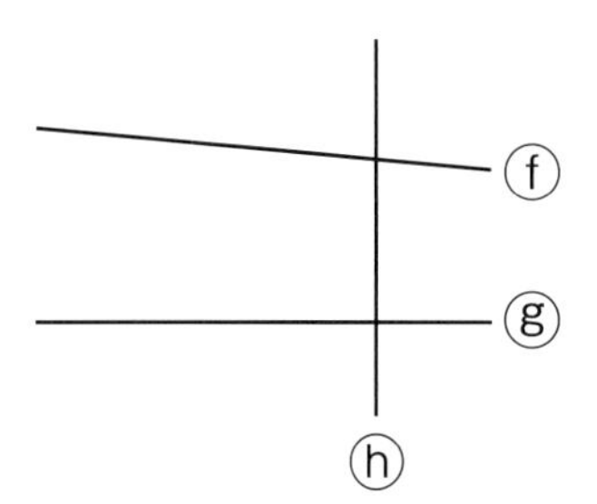

① As for straight lines ⓐ and ⓑ, how do they intersect straight line ⓒ?

Two straight lines are called **parallel** when a third straight line is perpendicular to both straight lines.

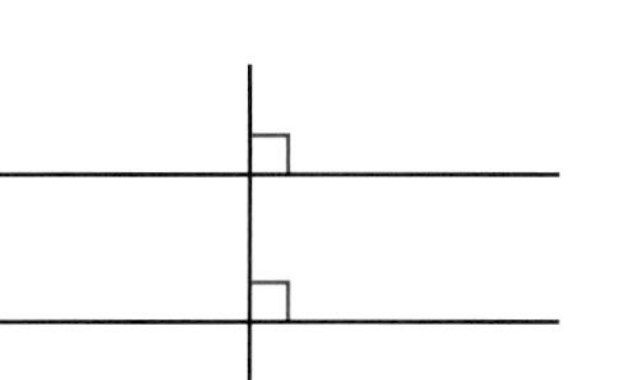

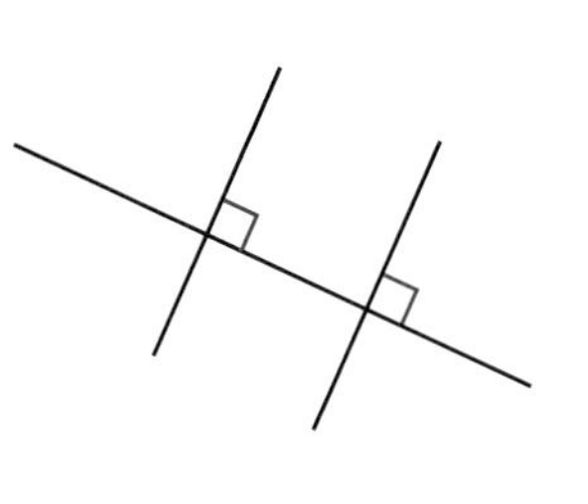

② Can you say that straight lines ⓕ and ⓖ are parallel? In the case they are not, let's explain the reason.

The above straight lines, ⓐ and ⓑ, are parallel straight lines.

Want to try

Are straight lines ⓚ and ⓛ parallel?

Let's extend straight line ⓚ and investigate.

The 2 straight lines on the right are also parallel.

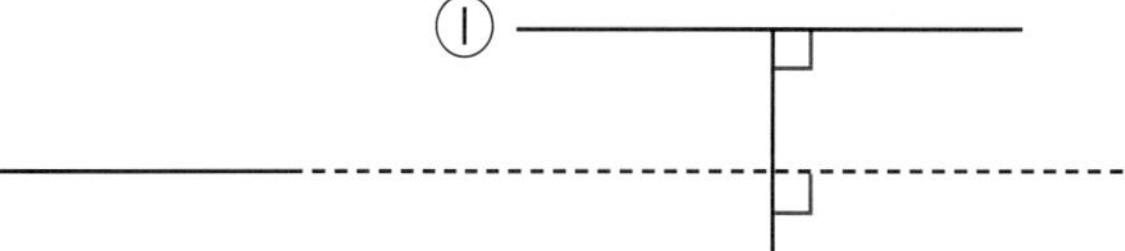

In the following figure, straight lines ⓐ and ⓑ are parallel. Let's explore the size of the angle formed by the intersection of the diagonal with the horizontal straight lines.

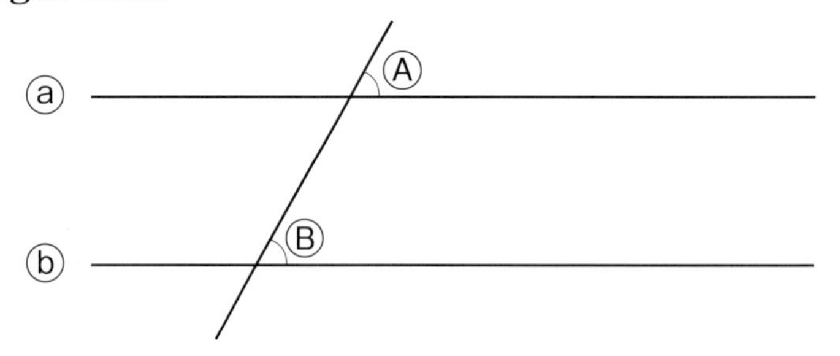

① How many degrees are angles Ⓐ and Ⓑ?

Parallel straight lines are intersected by another straight line at an equal angle.

② Let's draw a diagonal straight line on the above diagram and examine the angle formed by the intersection to straight lines ⓐ and ⓑ.

Want to confirm

In the following diagram, which straight lines are parallel?

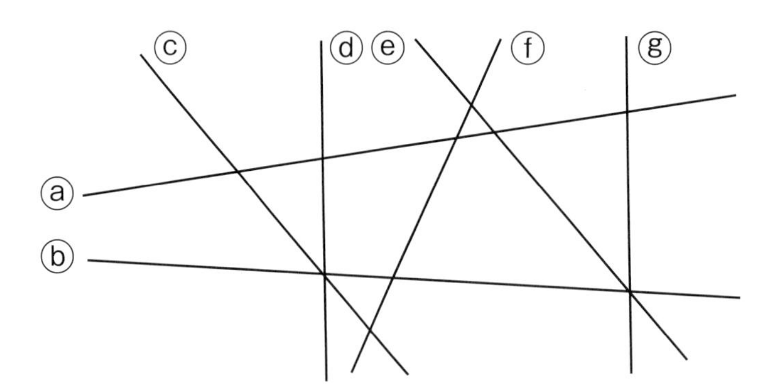

Want to try

In the following diagram, straight lines ⓕ and ⓖ are parallel. Let's find the size of angles Ⓕ～Ⓗ.

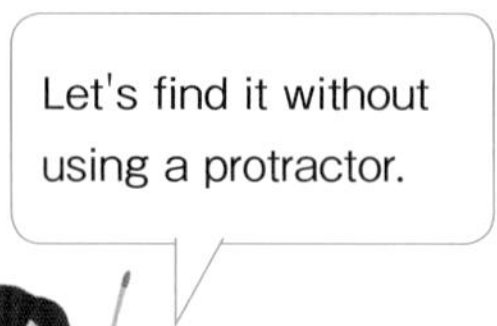

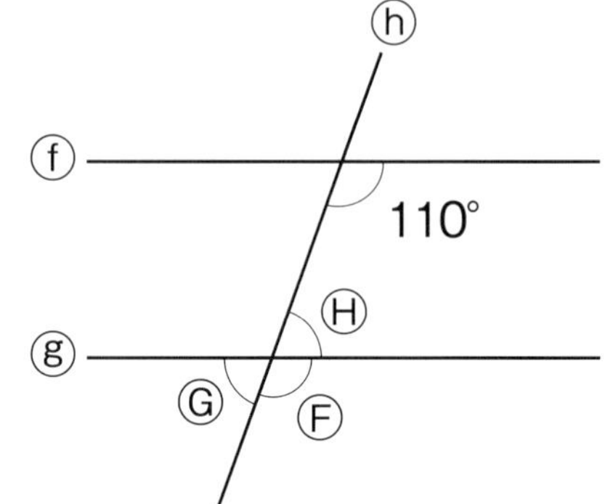

Want to compare Relationship of parallel straight lines

2 **In the following figure, straight lines ⓐ and ⓑ are parallel. Let's draw some perpendicular straight lines between the 2 straight lines, and compare their length.**

The length between 2 parallel straight lines is equal at every point. Also, both straight lines never cross, no matter how far they are extended.

Want to explain

5 Let's explain that the opposite sides, on a rectangle and square, are parallel.

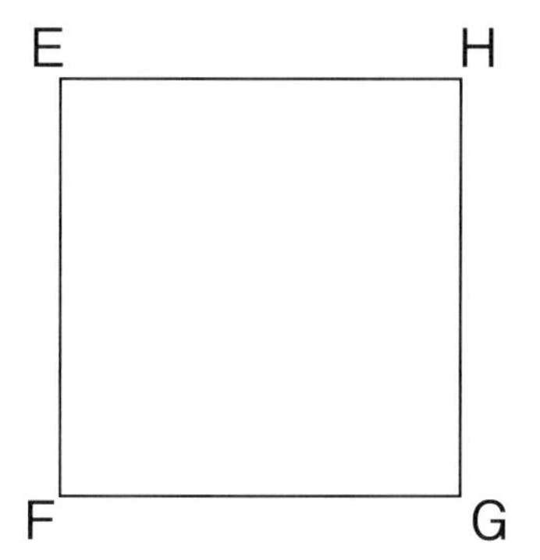

Want to find in our life

6 From our surroundings, let's look for perpendicular and parallel things.

Want to explain

Activity

3 **Let's think about how to draw a parallel straight line to straight line ⓐ. Let's read the drawing methods from Daiki and Yui below and explain the reason why both straight lines are parallel.**

ⓐ ———————————————

What properties are they using?

01402

Daiki's way of drawing

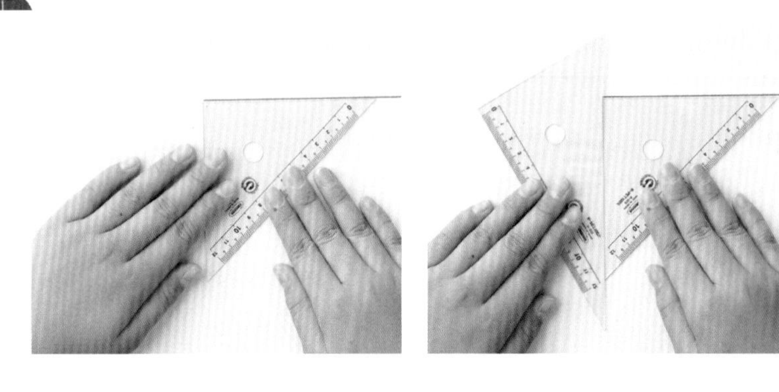

Yui's way of drawing

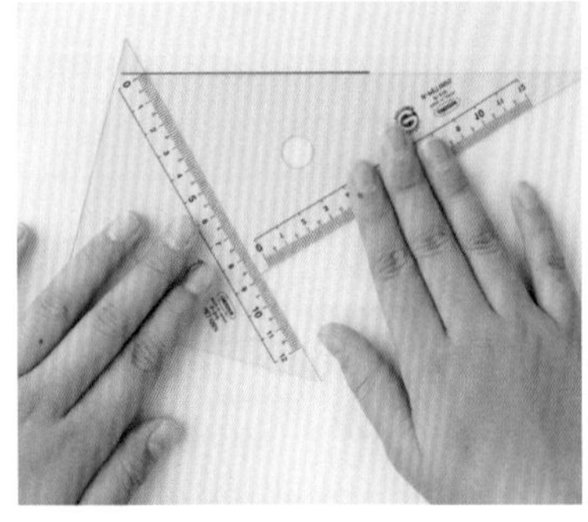

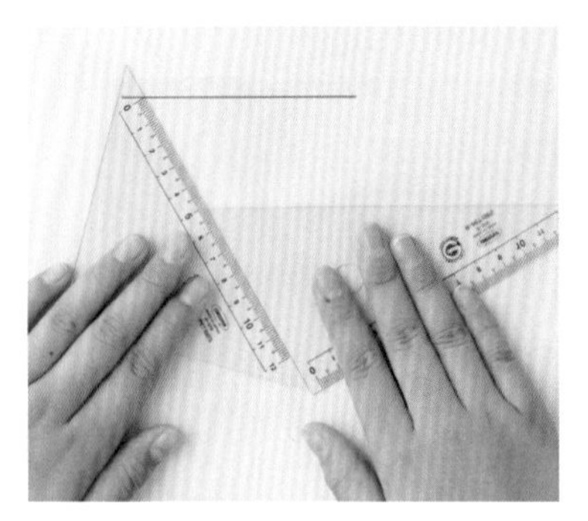

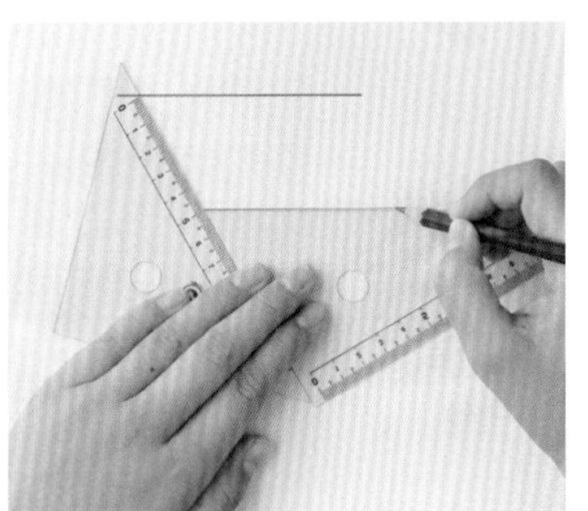

Let's draw a parallel straight line to straight line ⓑ through the point A.

ⓑ ————————————————————

• A

How to draw parallel straight lines

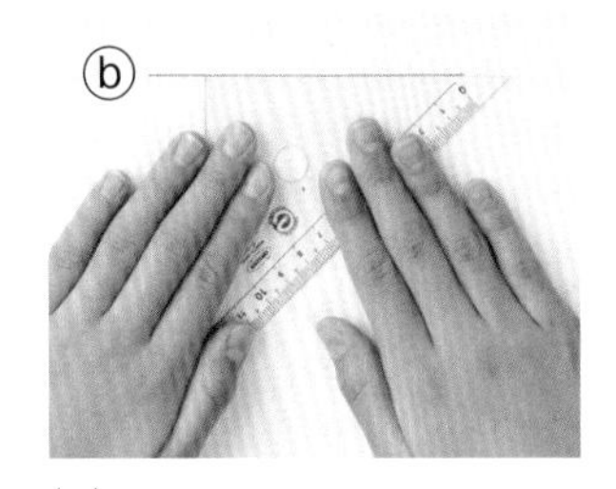

(1) Match the triangle ruler with straight line ⓑ.

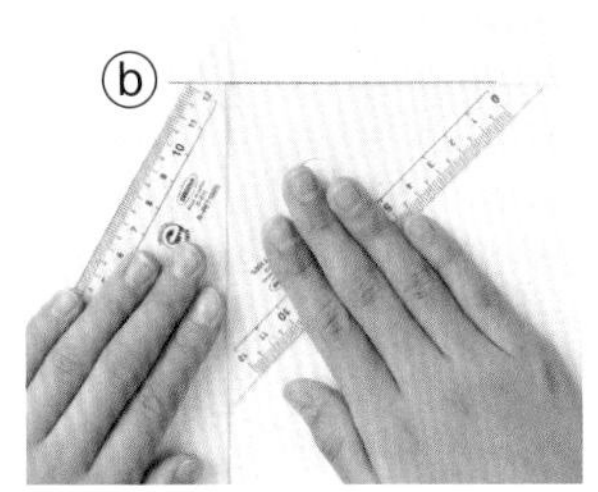

(2) Also, match the other triangle ruler.

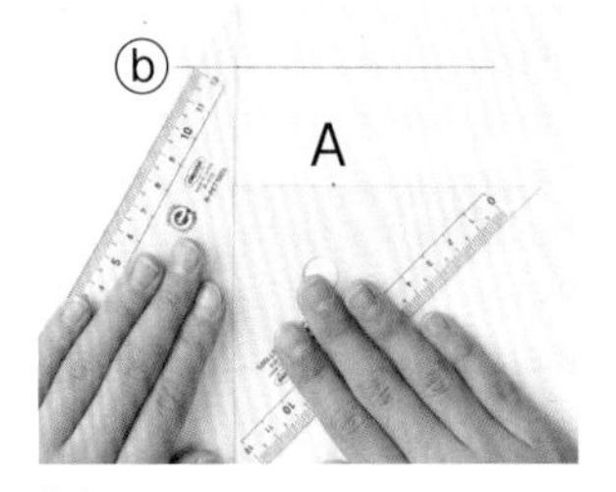

(3) Slide the right hand triangle ruler to match point A.

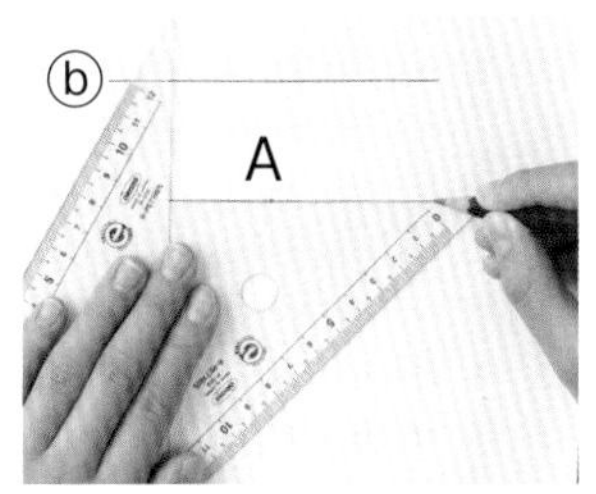

(4) Draw a straight line through point A while holding the triangle ruler.

Want to confirm

8 Let's draw the following straight lines on the diagram below.

① Straight line ⓓ that goes through point B and it is parallel to straight line ⓒ.

② Straight lines ⓔ and ⓕ, each 2 cm far from straight line ⓒ and parallel to it.

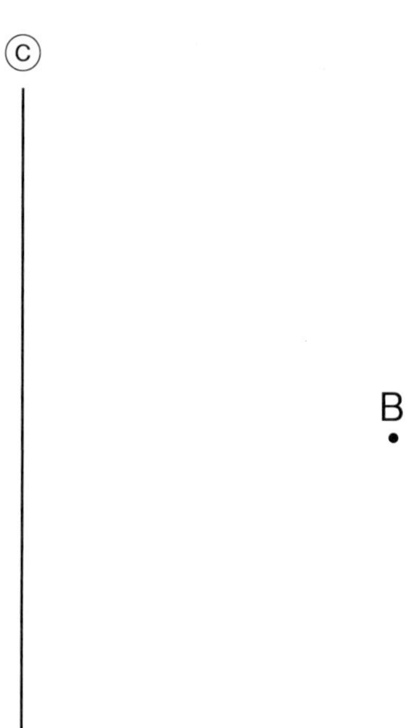

Want to try

9 Look at the following straight lines. Let's connect dots to draw parallel straight lines for each of them.

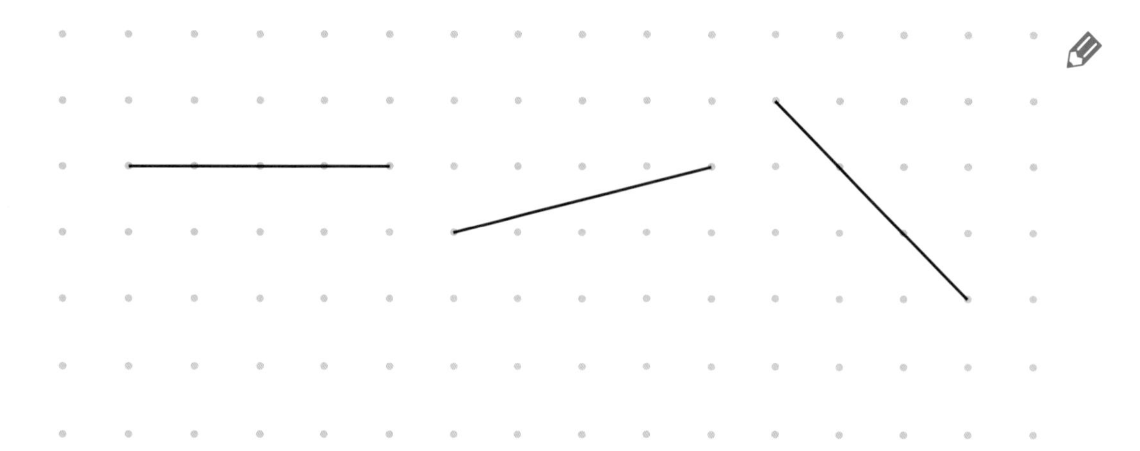

3 Various quadrilaterals

Want to explore Classification of quadrilaterals

1 **Let's make various quadrilaterals by connecting the dots of the cards. Also, let's examine the quadrilaterals that will be made.**

Let's use the cards on p. 161.

Ⓐ

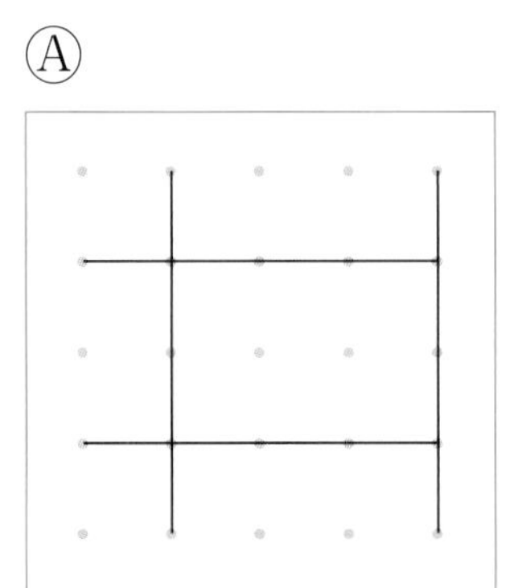

Ⓑ

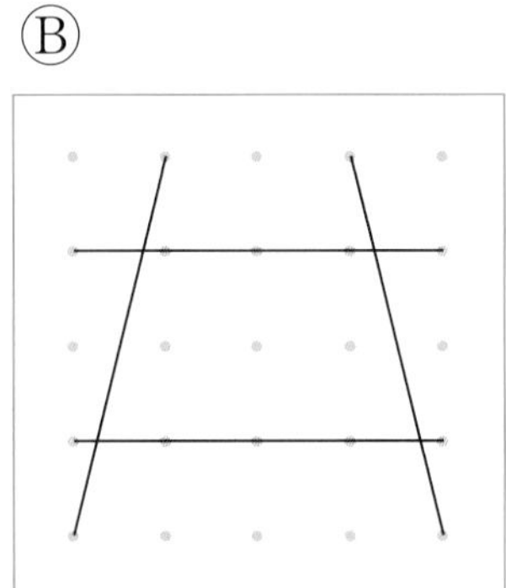

Ⓒ

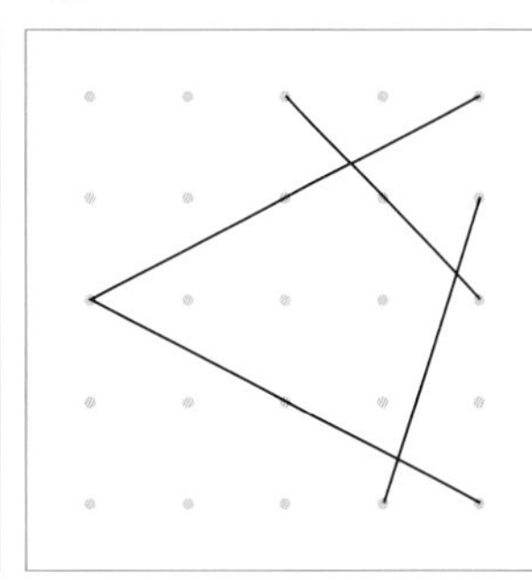

Ⓓ

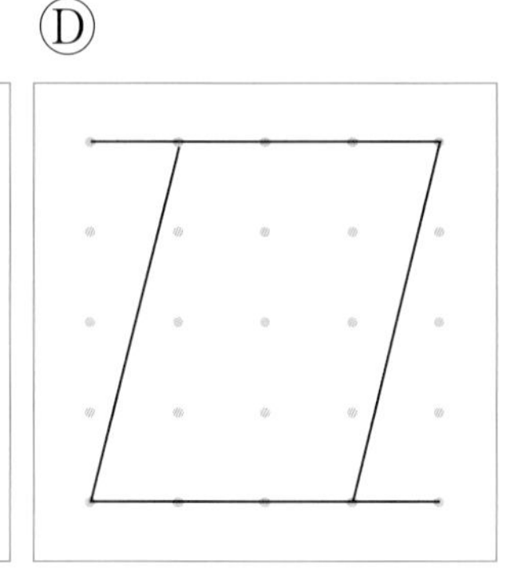

Ⓔ

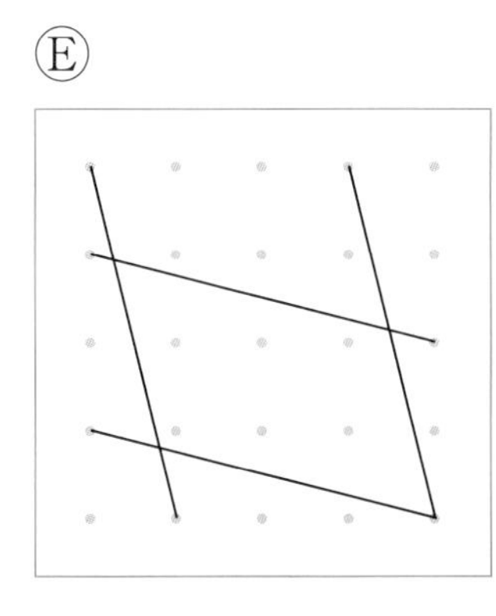

Ⓕ

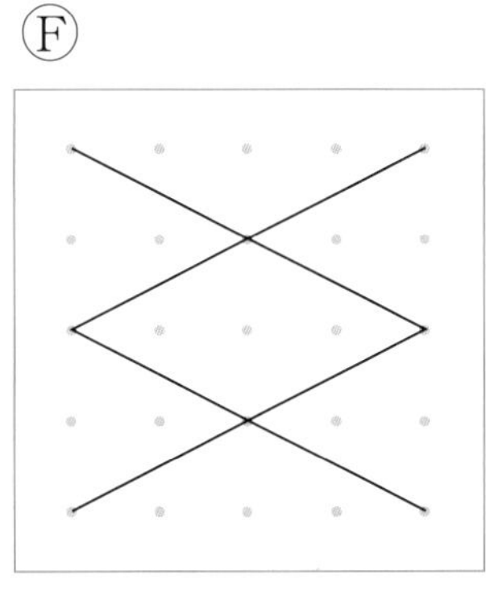

Ⓖ

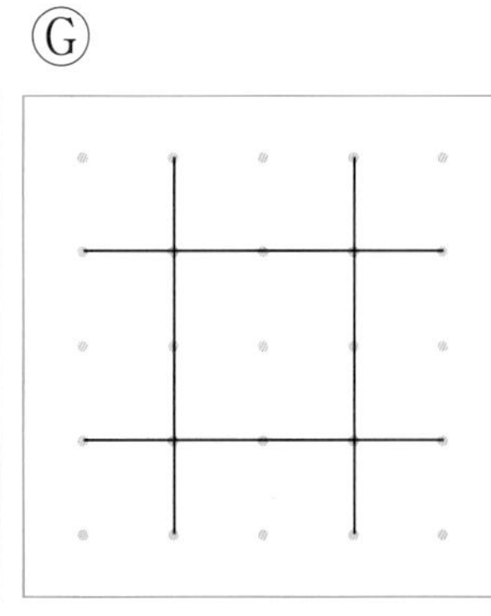

Ⓗ

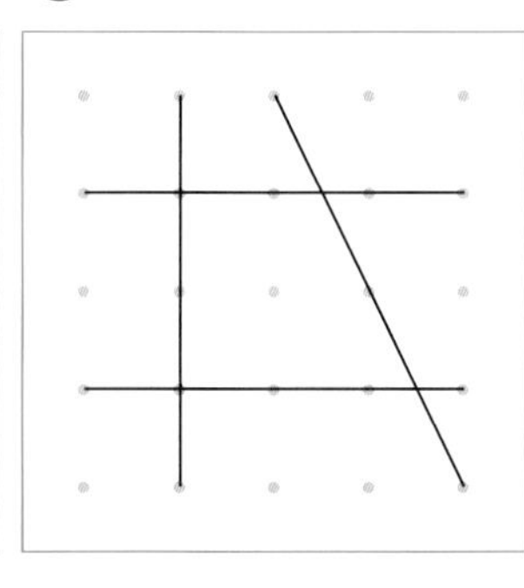

① Let's use colors to paint parallel sides.

② Let's identify the number of parallel straight line pairs to classify the above Ⓐ〜Ⓗ quadrilaterals into three groups.

One pair of sides are parallel.	Two pairs of sides are parallel.	No pair of parallel sides.
	 Ⓐ	

2 **Let's color the quadrilaterals that have one pair of opposite parallel sides on the right.**

Ⓑ

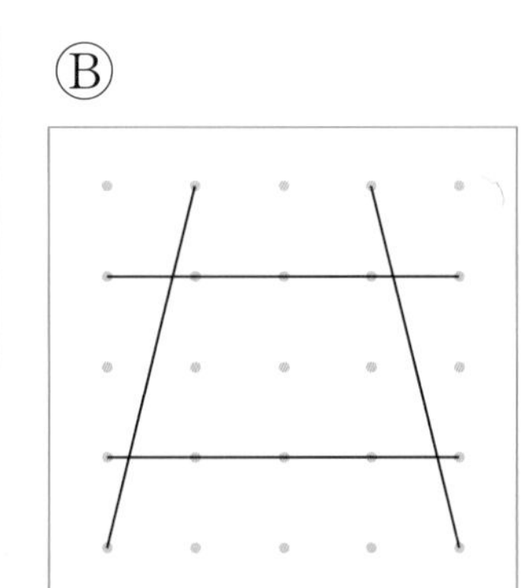

Ⓗ

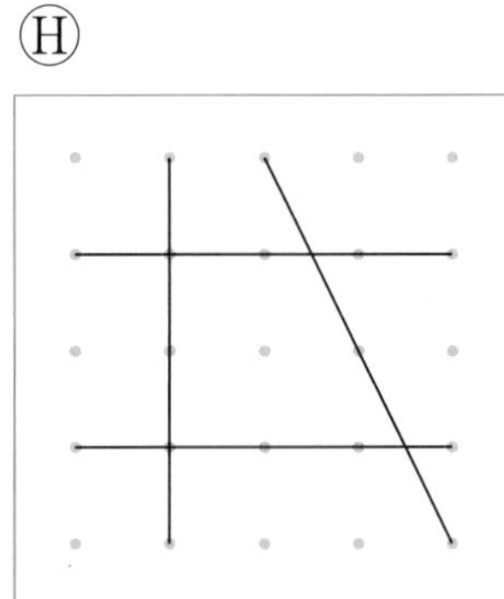

A quadrilateral that has one pair of opposite parallel sides is called a **trapezoid.**

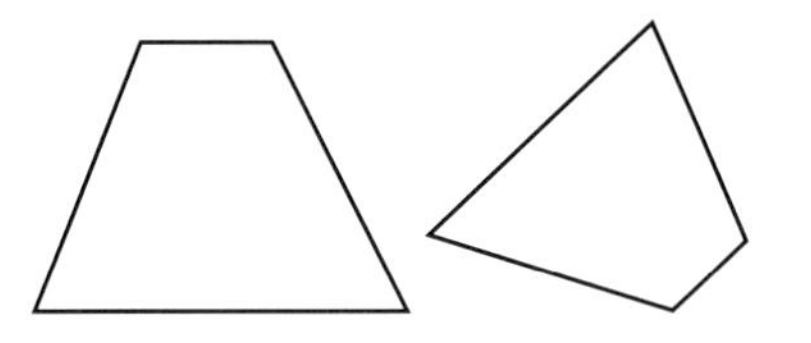

Want to try

1 Let's draw a trapezoid using 2 parallel straight lines.

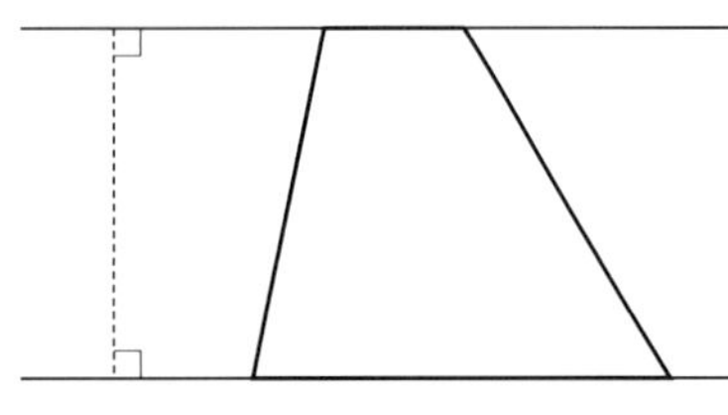

Want to find in our life

2 From your surroundings, let's look for things with the shape of trapezoids.

Want to know Parallelogram

Let's color the quadrilaterals that have two pair of opposite parallel sides on the right.

Ⓓ

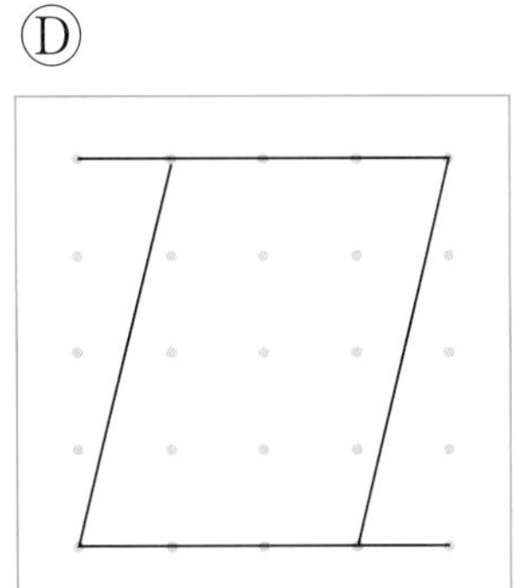

Ⓔ

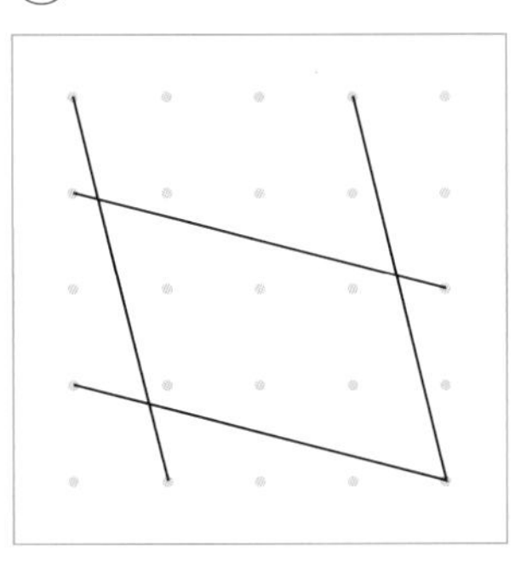

A quadrilateral that has two pair of opposite parallel sides is called a **parallelogram**.

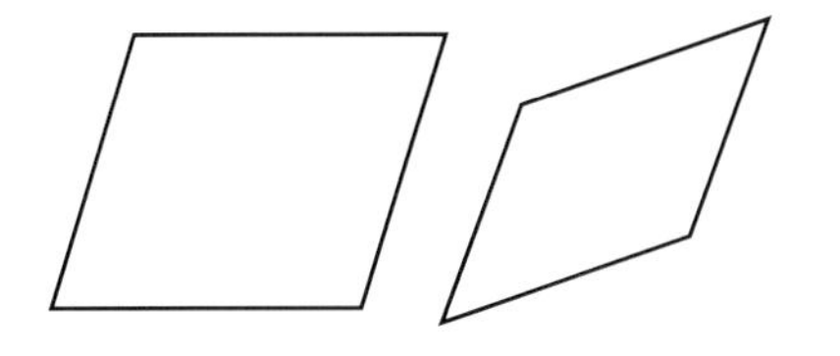

From the group with two opposite parallel sides, Ⓐ is a rectangle and Ⓖ is a square.

Want to try

Let's draw parallelograms using the following grid paper.

Want to find in our life

From your surroundings, let's look for things with the shape of parallelograms.

Numazu-gaki, Numazu City, Shizuoka Prefecture.

Want to explain

4 **Let's find trapezoids and parallelograms from the following quadrilaterals. Also, let's explain the reasons why they become trapezoids or parallelograms.**

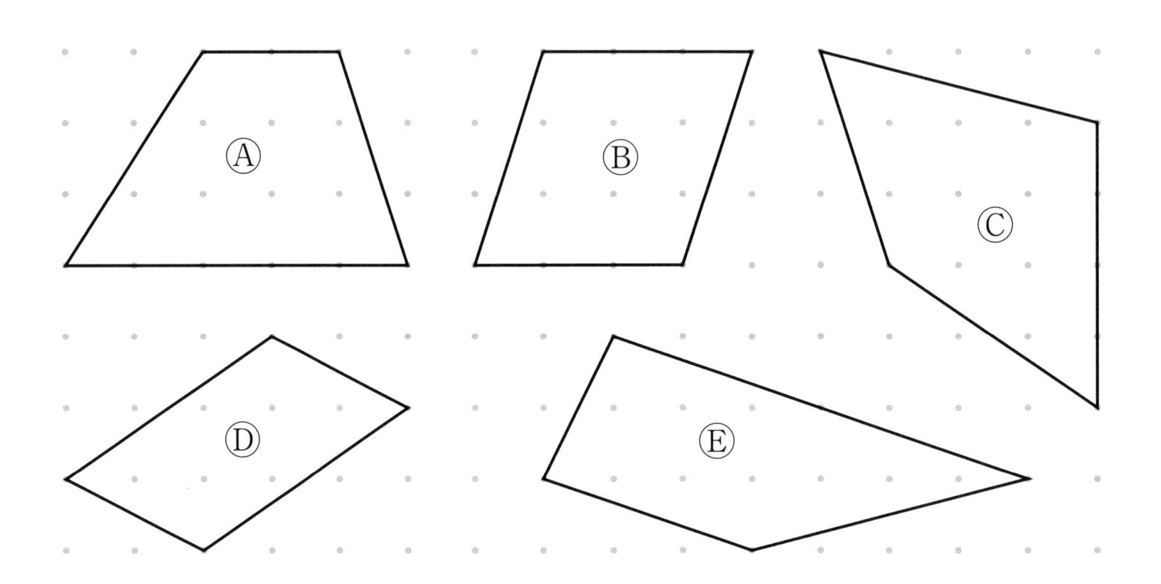

Yui's idea

As shown on the right diagram, each pair of opposite sides of the figure Ⓐ were colored.
The red pair of opposite sides are parallel.
The blue pair of opposite sides are not parallel.
Therefore, this figure is a trapezoid.

For the other figures, let's also write down the reasons on your notebook.

Want to explore Properties of parallelograms

Let's explore the length of the opposite sides and the size of the opposite angles in a parallelogram.

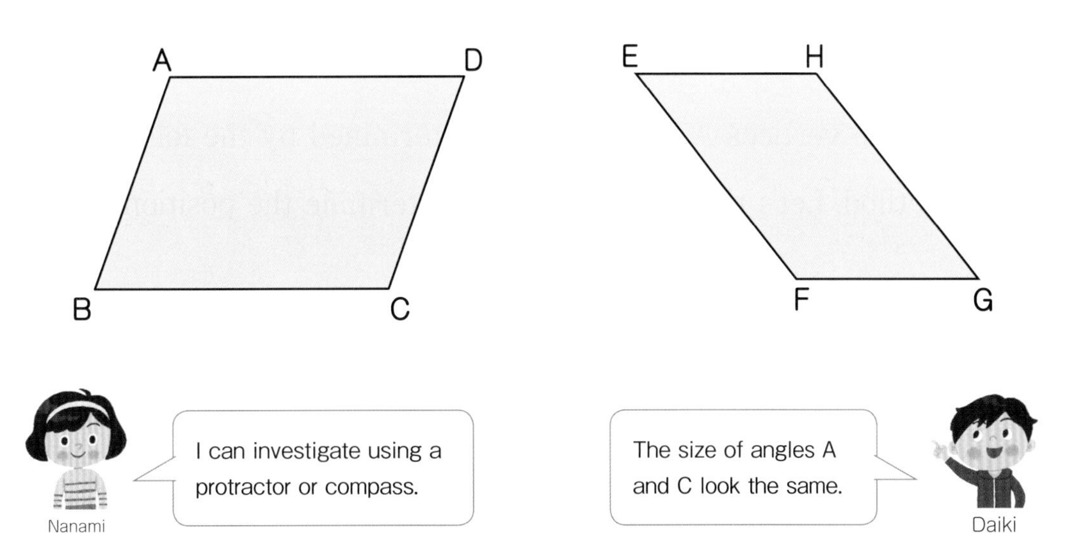

In parallelograms, the lengths of the opposite sides are equal and the size of opposite angles are equal.

Want to confirm

What is the length (cm) of side AD and side CD shown on the right parallelogram? Also, what is the size in degrees of angles C and D?

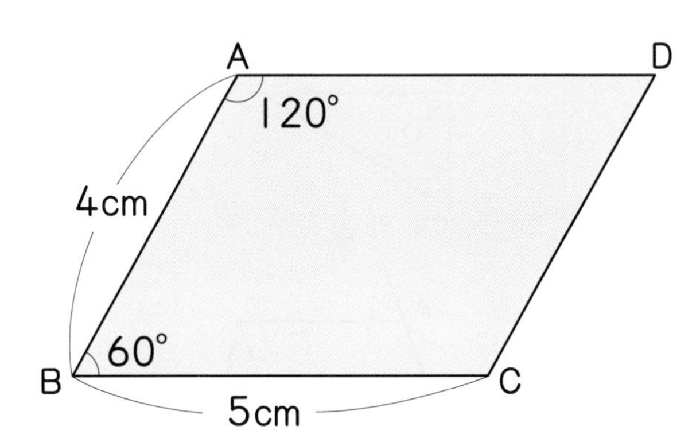

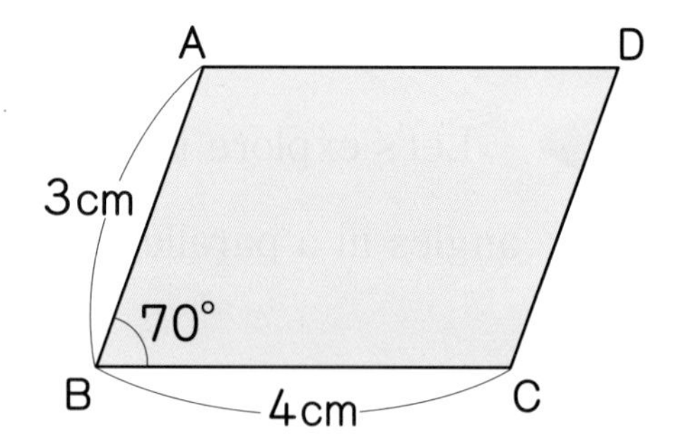

5 Let's think about how to draw a parallelogram like the one shown on the right.

① The vertices A, B, and C are determined by the following drawing method. Let's think about how to determine the position of vertex D.

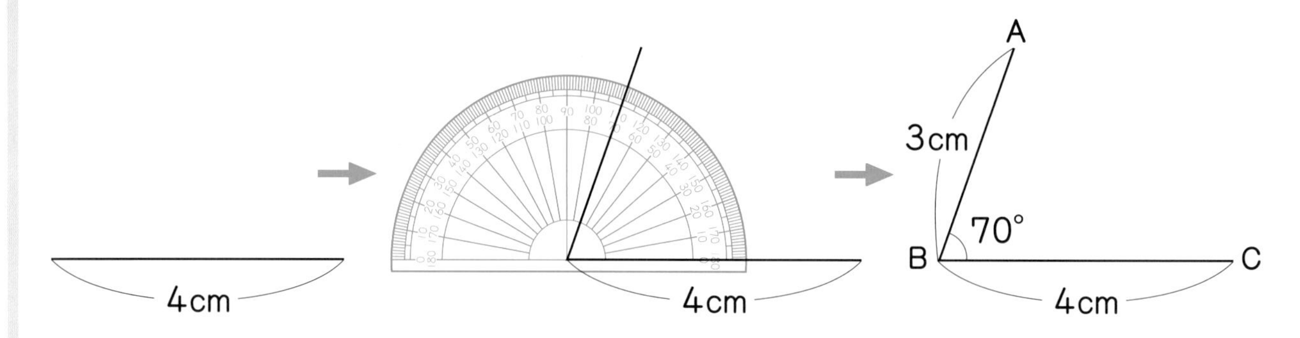

② Let's explain the idea of each child.

Hiroto's idea

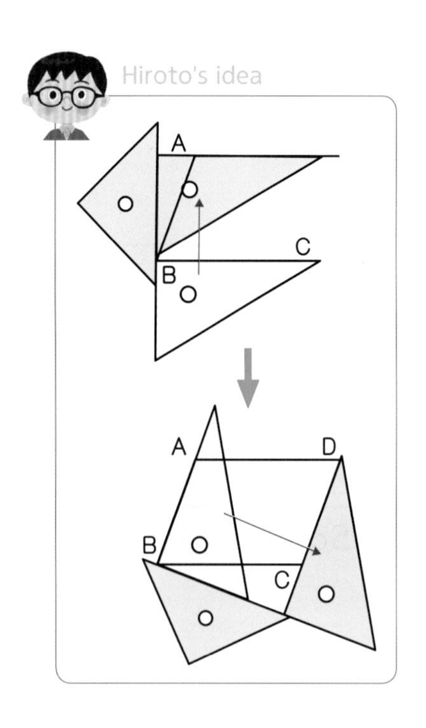

Yui's idea

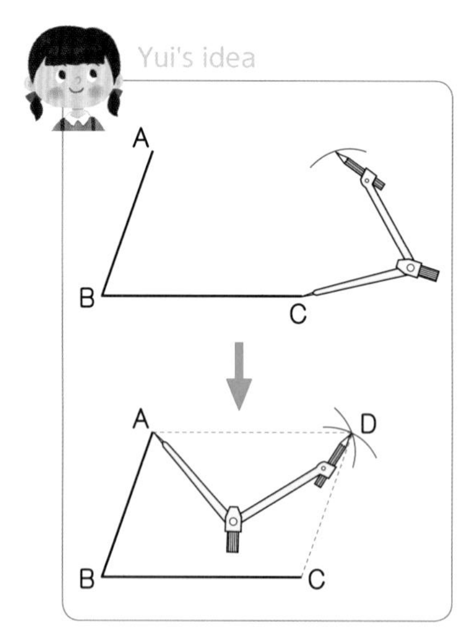

Nanami's idea

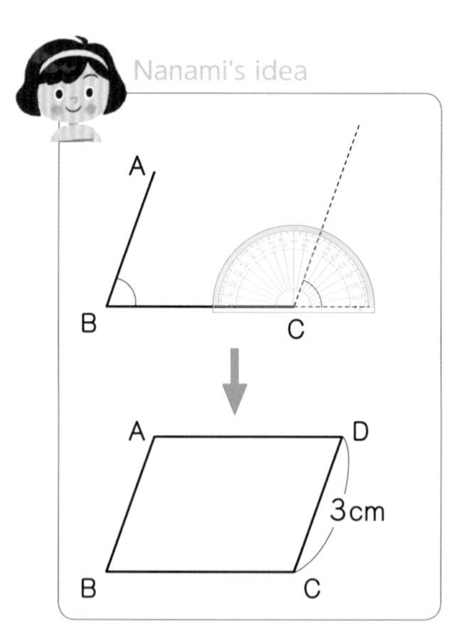

Which properties of parallelograms are they using?

7 Let's draw the following parallelograms.

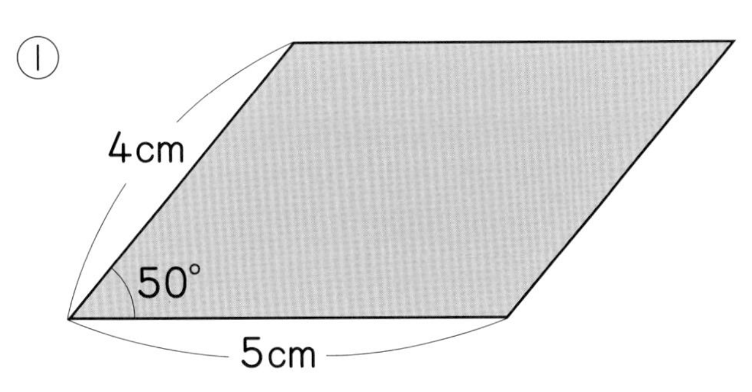

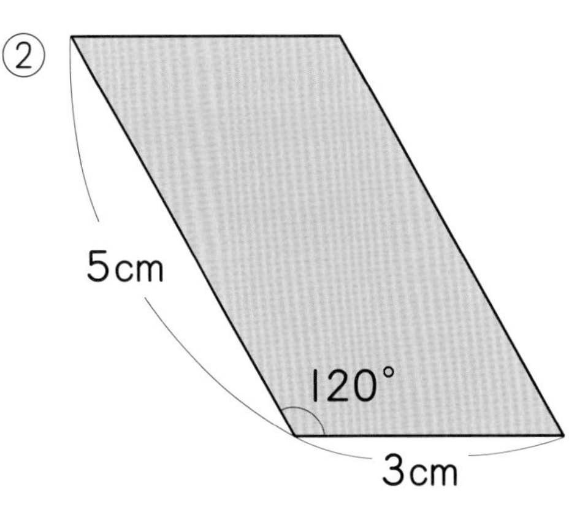

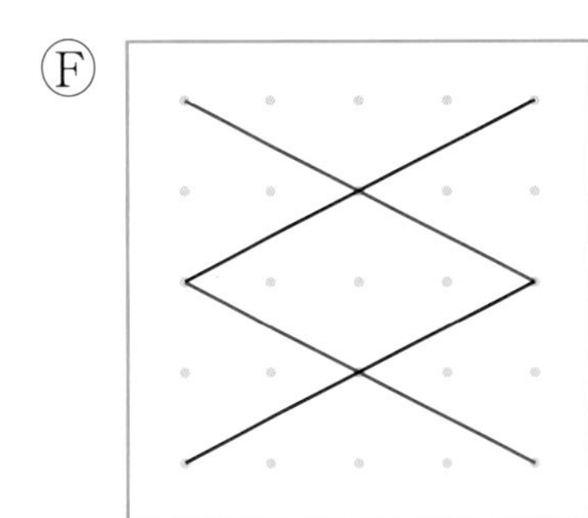

6 **Let's examine the length of the four sides of quadrilateral Ⓕ from page 85.**

A quadrilateral with four equal sides is called a **rhombus**.

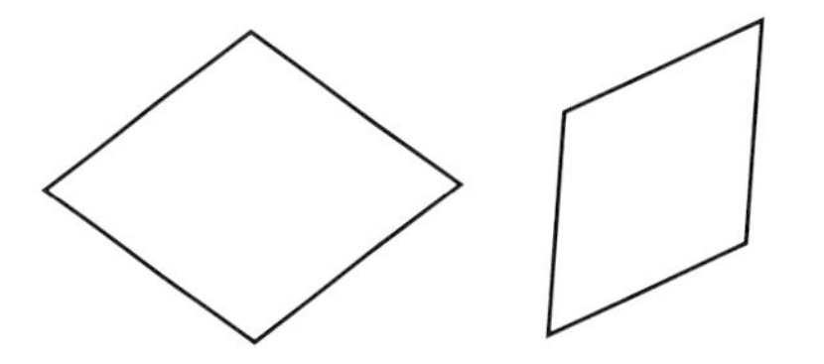

Want to confirm

8 The diagram below shows two parts of perimeters of circles centered at A and C, and with the same radius. They intersect at B and D. Let's explore this diagram.

① Let's connect the points A→B→C→D→A with straight lines to draw a quadrilateral.

② Let's examine the lengths of the sides and size of the angles. Which quadrilateral is this?

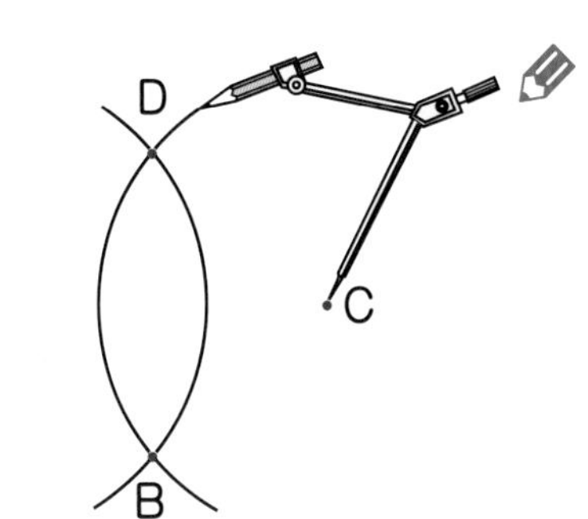

Want to explore

Let's explore the rhombus that you drew on exercise 8 from the previous page.

① Are the sizes of opposite angles equal?

② Are the opposite sides parallel?

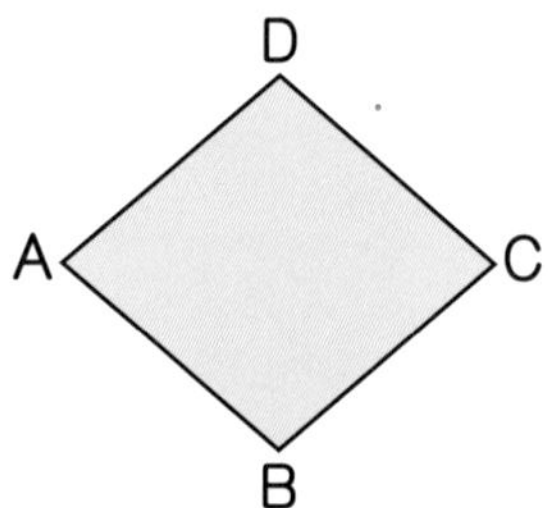

In a rhombus, the size of opposite angles are equal and the opposite sides are parallel.

Want to know

What is the length (cm) of sides AB, BC, and CD shown in the right rhombus?

Also, what is the size in degrees of angles A and D?

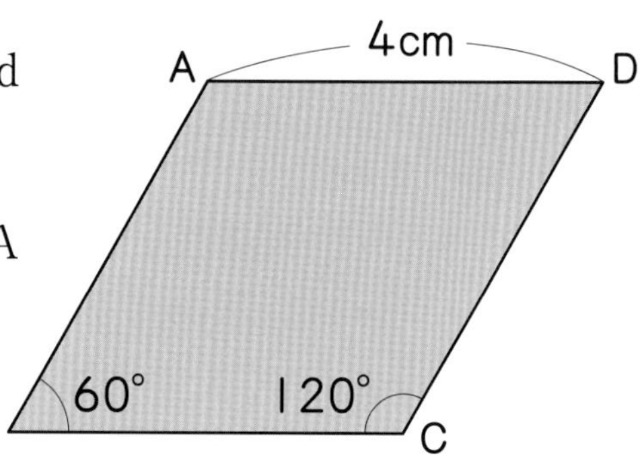

Want to confirm

Let's think about how to draw the rhombus shown on the right.

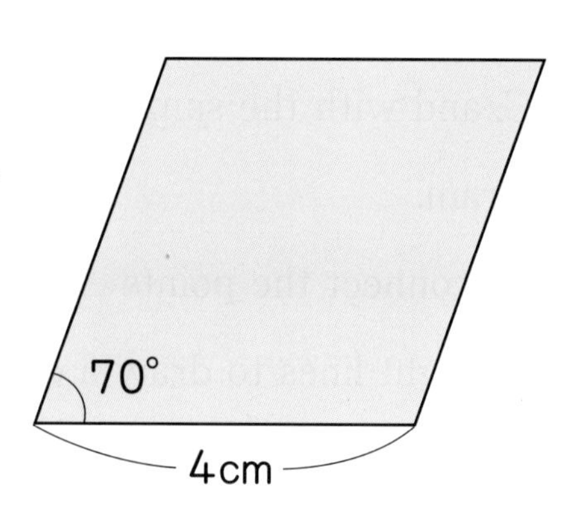

Way to see and think

All four sides of the rhombus have the same length.

4 Diagonals of quadrilaterals

Want to discuss

1 **Let's connect the opposite vertices of the quadrilaterals with straight lines. Also, let's discuss what you noticed.**

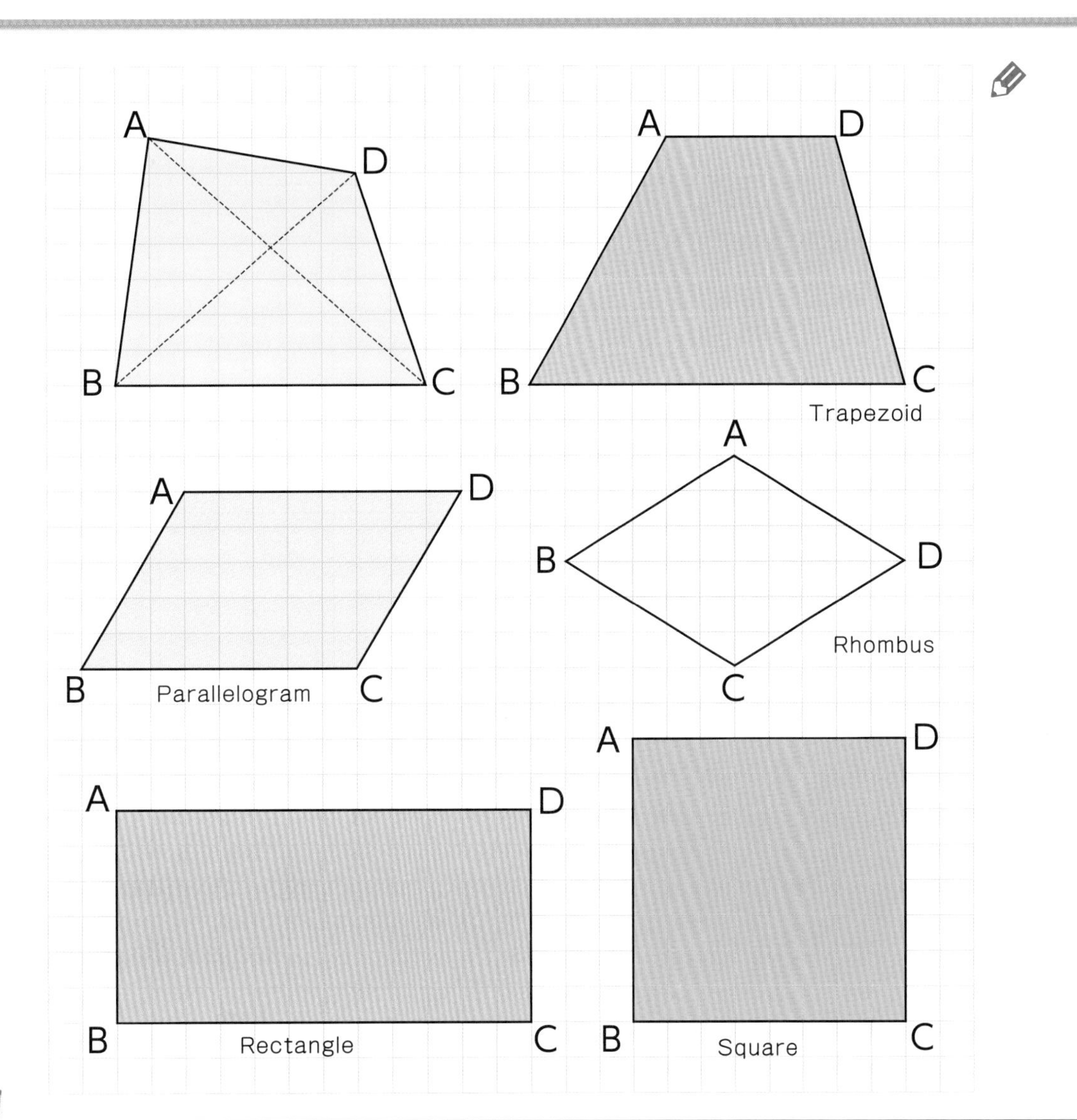

Each straight line that you drew to connect the opposite vertices is called a **diagonal.** There are two diagonals in each quadrilateral.

What is going on between the length of the diagonals and the intersecting angle?

There are figures with the same properties.

Want to solve

1 Let's summarize the properties of the quadrilateral's diagonals that you found in exercise **1** on the previous page in the table below. Write a circle ○ on what always applies.

Properties of the quadrilateral's diagonals \ Quadrilateral's name	Trapezoid	Parallelogram	Rhombus	Rectangle	Square
① Both diagonals have equal length.					○
② Both diagonals intersect at their midpoint.					○
③ Both diagonals are perpendicular lines.					○

Want to confirm

Draw the following quadrilaterals using the properties investigated in **1**.

① A rhombus with 4 cm and 3 cm length diagonals.

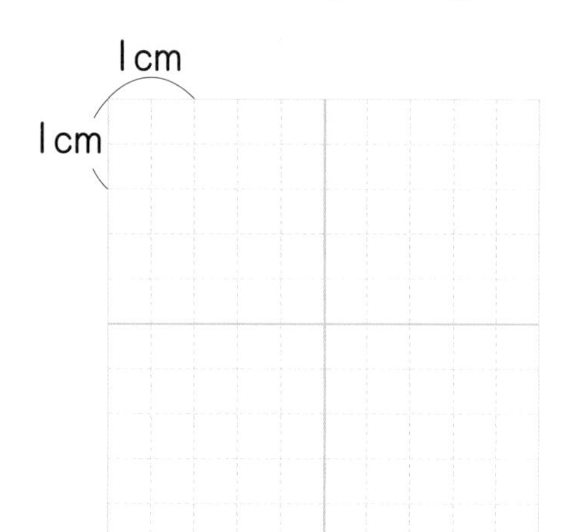

② A square with 4 cm length diagonals.

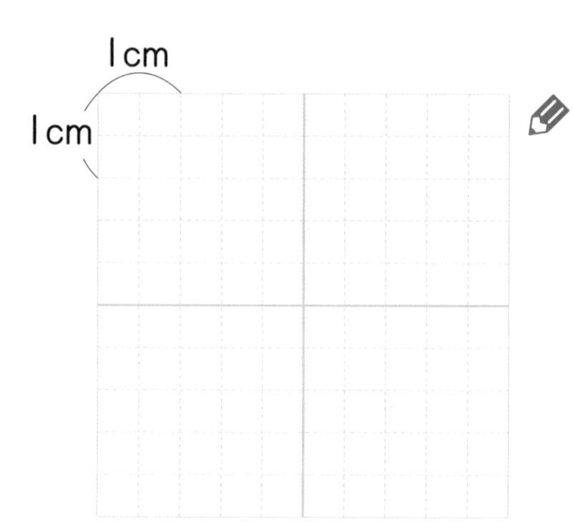

5 Relationship of quadrilaterals

Want to explore

1 **Let's draw a parallelogram with a 4 cm and 6 cm length sides.**

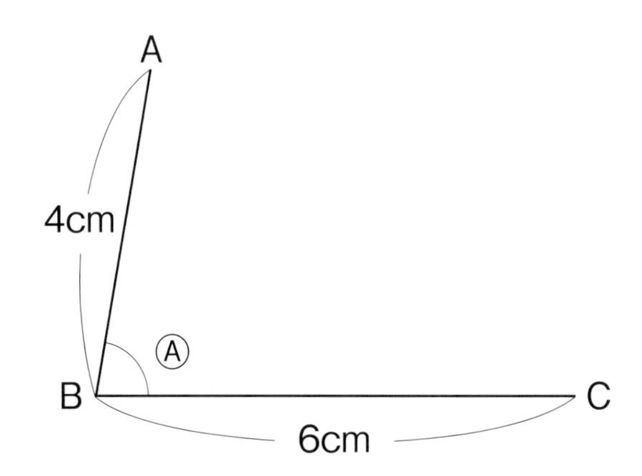

① Let's draw angle Ⓐ with sizes 80° and 120°.

② Let's draw angle Ⓐ with size 90°. What quadrilateral can be drawn?

Want to try

Let's draw, as the following diagram, a rhombus with a 5 cm length side.

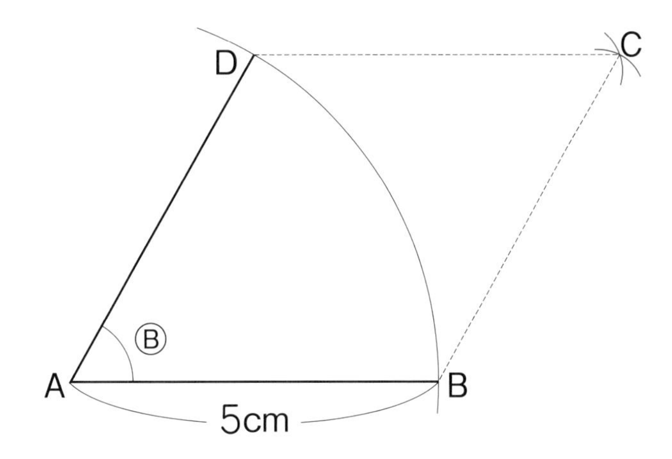

① Let's draw angle Ⓑ with size 60°.

② Let's draw angle Ⓑ with size 120°.

③ Let's draw angle Ⓑ with size 90°.

What quadrilateral can be drawn?

What is the size in degrees for the 4 angles in each drawing?

Want to discuss

Let's discuss what you have learned so far on quadrilaterals, based on the properties you found in page 94 and noticed things from page 95. Let's talk about the similarities and differences.

Develop in Junior High School

Relationships of Quadrilaterals

The following classification of various quadrilaterals learned so far can be made based on the length of sides and the size of the angles.

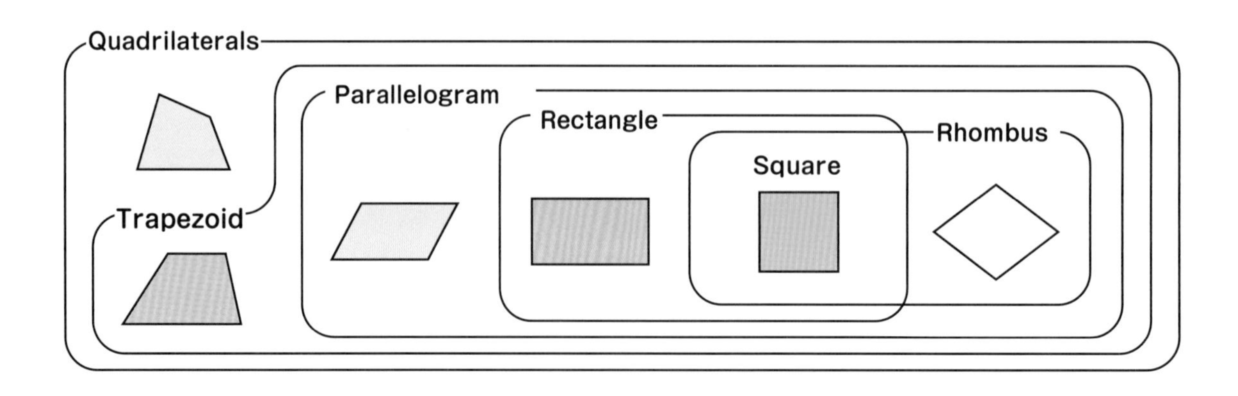

6 Patterns by tessellation

Want to explore

1 **Use parallelograms, rhombuses, and trapezoids that have the same shape and size to make tessellation patterns. Let's color the shapes.**

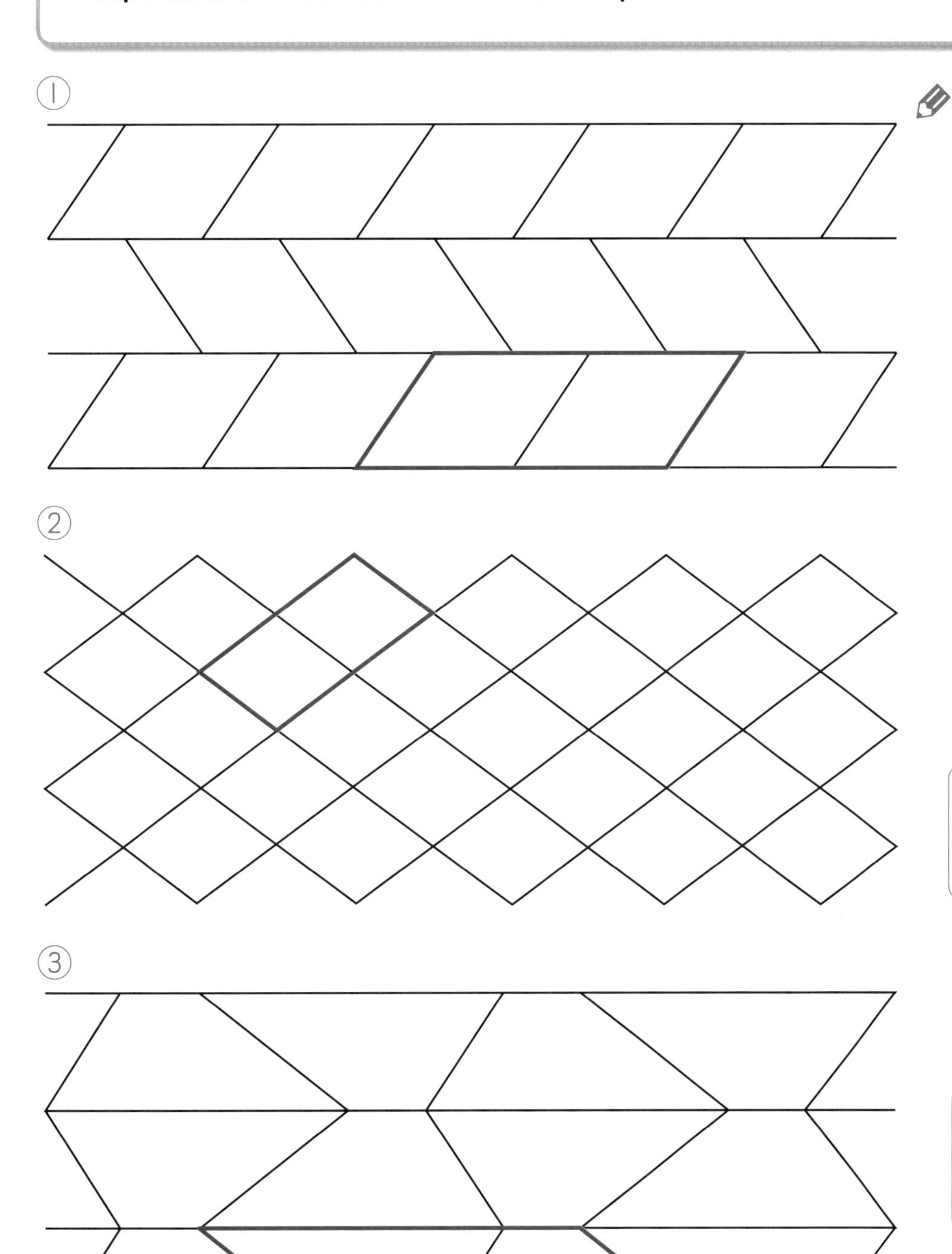

Even two parallelograms together form a parallelogram.

Inside the rhombus pattern I found a parallelogram.

Also, inside the trapezoid pattern I found a parallelogram.

Why all of them became a parallelogram?

Want to find in our life

From your surroundings, let's find tessellation patterns.

Sidewalk of Jimbocho Station (Chiyoda City, Tokyo)

Arrow-Feather design (Hamamatsu City, Shizuoka Prefecture)

Namako Wall (Matsuzaki Town, Kamo-gun, Shizuoka Prefecture)

Checker flag

What you can do now

Understanding the relationship between perpendicular and parallel straight lines.

1 Passing through point A, let's draw perpendicular and parallel straight lines to straight line ⓐ.

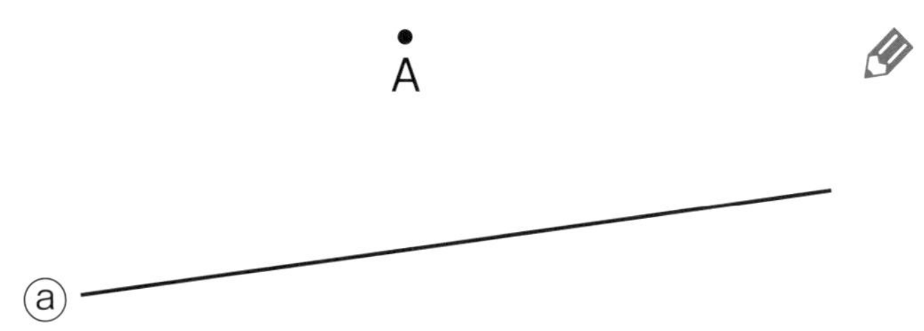

Understanding the properties of various quadrilaterals.

2 Let's answer the following questions.

① Look at the figures on the right. Let's fill in the ☐ with appropriate words.

ⓐ A quadrilateral that has one pair of ☐ opposite sides is called a ☐.

ⓑ A quadrilateral that has two pairs of ☐ opposite sides is called a ☐.

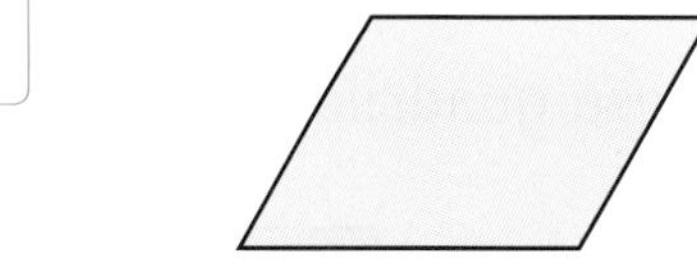

ⓒ A quadrilateral that has 4 sides with ☐ length is called ☐.

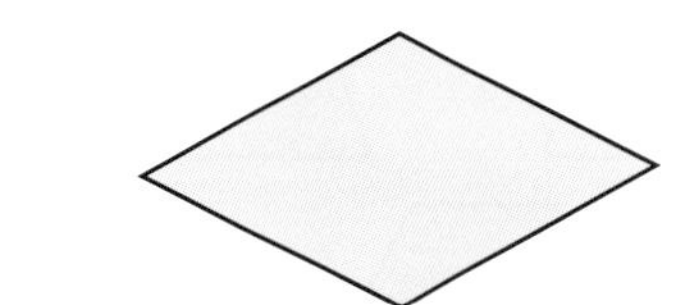

② Let's draw a rhombus with 5 cm and 3 cm length diagonals.

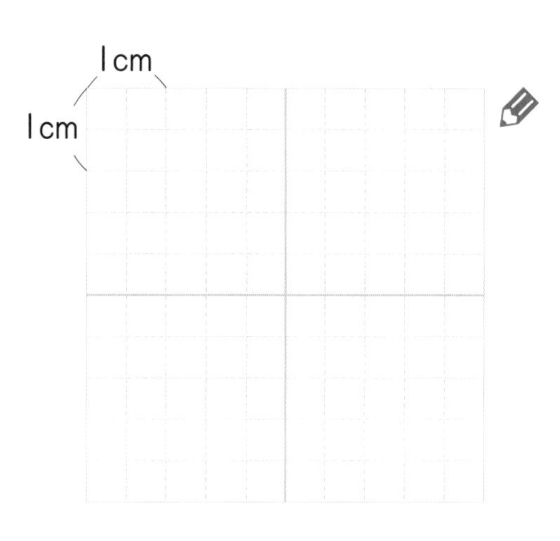

Supplementary problems p. 153

Usefulness and efficiency of learning

1 In the figure on the right, let's identify two perpendicular lines. Also, let's identify parallel lines.

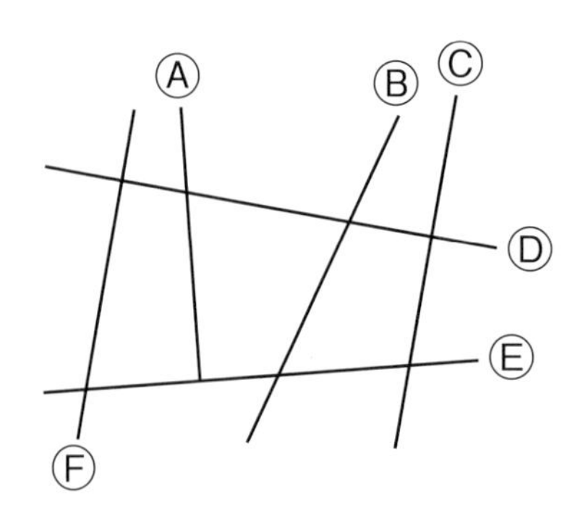

☐ Understanding the relationship between perpendicular and parallel straight lines.

2 There is a parallelogram as shown on the right. Let's fill in the ☐ with numbers.

Also, let's draw the same parallelogram as shown on the right.

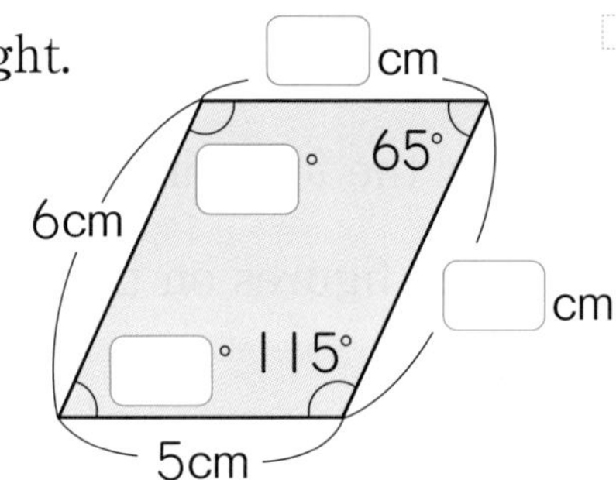

☐ Understanding the properties of various quadrilaterals.

3 Which of these quadrilaterals have the following properties?

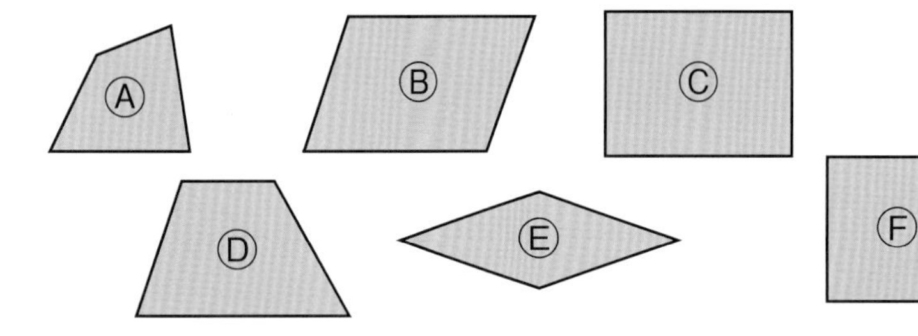

① Two pairs of parallel sides.
② Four angles with equal size.
③ Diagonals with equal length.
④ Opposite sides with equal length.
⑤ Opposite angles with equal size.
⑥ No parallel sides.

☐ Understanding the properties of various quadrilaterals.

Let's deepen.

Can you explain well why this happens to various quadrilaterals?

What kind of quadrilateral can you make?

Want to think

Let's think about the diagram on the right. What quadrilateral is created when you connect with straight lines the following four points in order?

Let's think and draw on the diagram below.

Also, let's discuss the reason for each process.

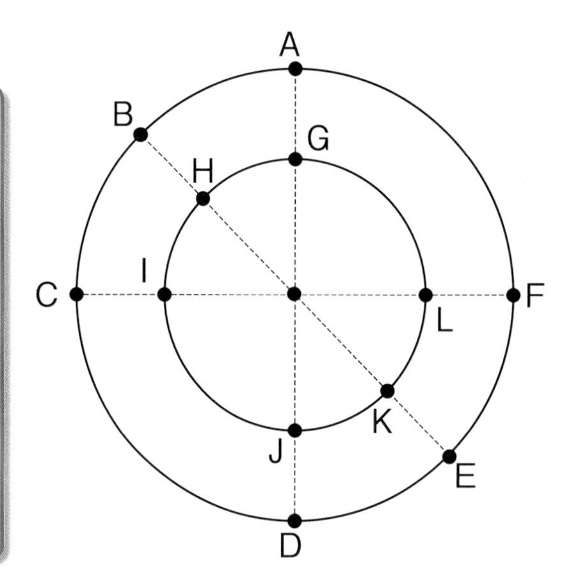

① point B, point C, point E, point F
② point G, point I, point J, point L
③ point G, point C, point J, point F
④ point A, point H, point D, point K

①

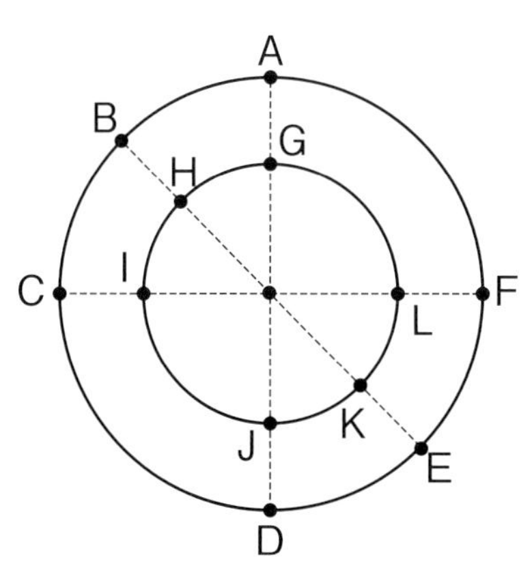

②

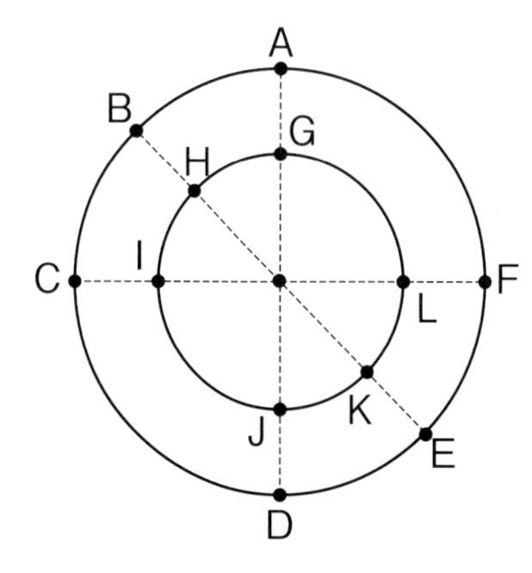

③

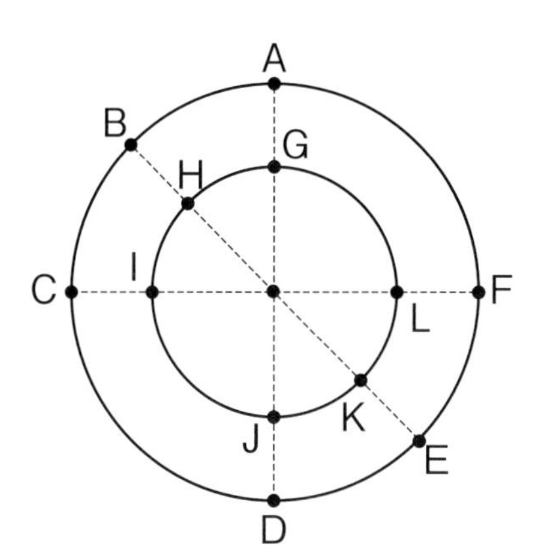

④

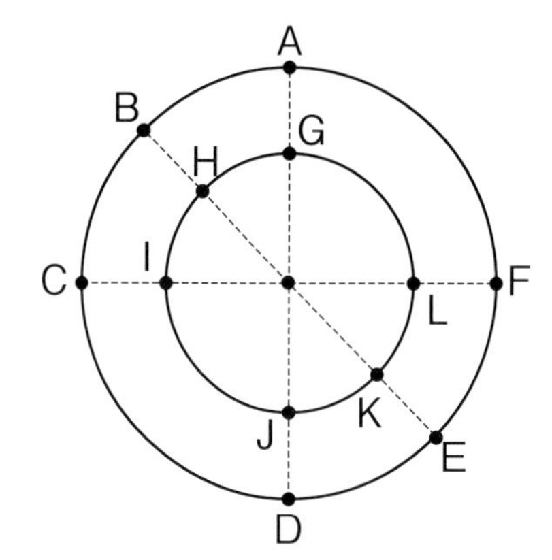

Division by 2-digit numbers?

How can you calculate a division by a 2-digit number?

Division by 2-digit number

Let's think about how to calculate in vertical form.

1 Division by tens

Want to solve

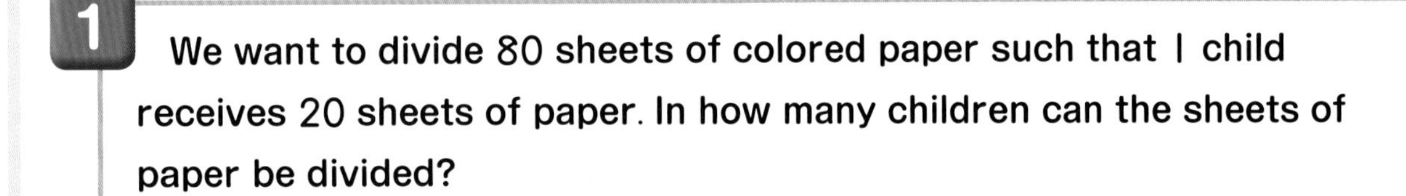

1 **We want to divide 80 sheets of colored paper such that 1 child receives 20 sheets of paper. In how many children can the sheets of paper be divided?**

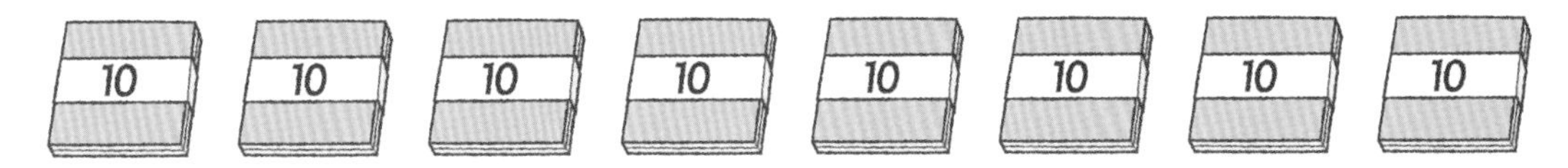

① Let's write a math expression. [ ]

② Let's think about how to calculate.

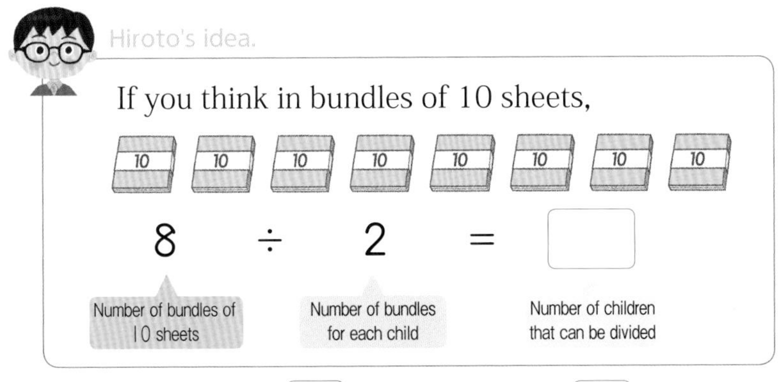

$80 \div 20 =$ [ ] Answer: [ ] children

As for $80 \div 20$, if you think based on 10, you can find the answer by calculating $8 \div 2$.

Want to confirm

Let's solve the following calculations.

① $40 \div 20$ ② $60 \div 30$ ③ $90 \div 30$ ④ $80 \div 10$

⑤ $150 \div 50$ ⑥ $120 \div 20$ ⑦ $240 \div 60$ ⑧ $160 \div 80$

Want to solve

2

We want to divide 140 sheets of colored paper such that 1 child receives 30 sheets of paper. In how many children can the sheets of paper be divided? How many sheets will remain?

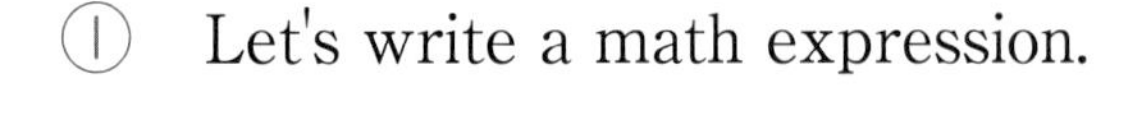

① Let's write a math expression.

② Let's think about how to calculate.

Daiki

If I think in groups of 10 sheets,
14 ÷ 3 = 4 remainder 2
Is the remainder 2 sheets?

The remainder is 2 bundles of 10, therefore ....

Nanami

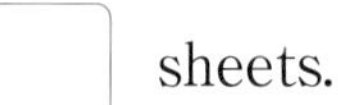

Answer: ☐ children can receive and the remainder is ☐ sheets.

③ Let's confirm the answer.

30 × ☐ + ☐ = ☐

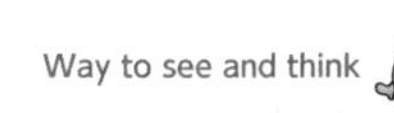

Divisor × Quotient + Remainder = Dividend

Want to confirm

Let's solve the following calculations.

① 70 ÷ 20 ② 80 ÷ 30 ③ 90 ÷ 40 ④ 50 ÷ 20

⑤ 320 ÷ 60 ⑥ 180 ÷ 70 ⑦ 200 ÷ 30 ⑧ 250 ÷ 80

2 Division by 2-digit number (1)

Want to solve The case: (2-digit) ÷ (2-digit)

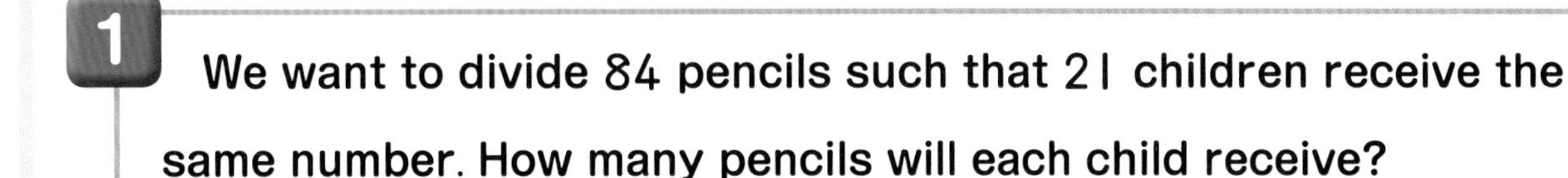

1 We want to divide 84 pencils such that 21 children receive the same number. How many pencils will each child receive?

① Let's write a math expression. [ ]

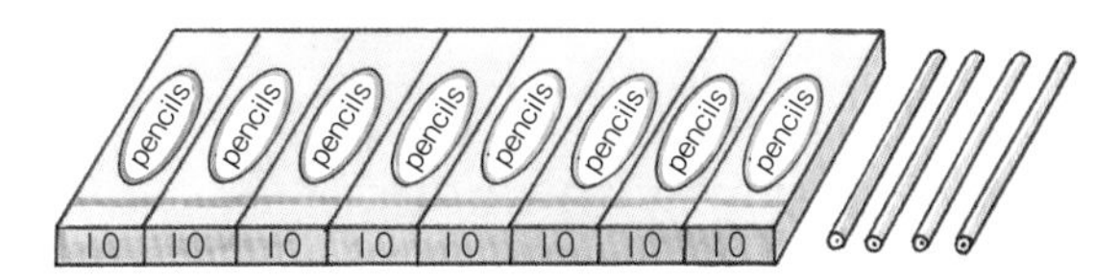

② If you calculate in vertical form, from which place value is the quotient written?

21) 84

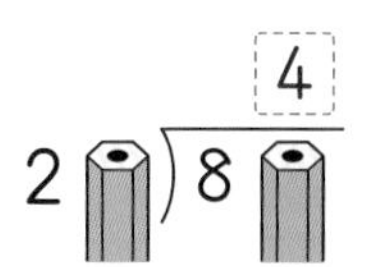

③ Think of 80 ÷ 20 and let's guess the quotient by 8 ÷ 2.

④ Is the quotient 4 correct? Let's confirm.

Division algorithm to calculate 84 ÷ 21 in vertical form

21) 84 → 2) 8 (quotient 4) → 21) 84, 84 (quotient 4) → 21) 84, 84, 0 (quotient 4)

From which place value → Divide → Multiply → Subtract

Want to confirm

Let's solve the following divisions in vertical form.

① 99 ÷ 33 ② 63 ÷ 21 ③ 64 ÷ 32

④ 48 ÷ 23 ⑤ 97 ÷ 32 ⑥ 91 ÷ 44

Want to think **Tentative quotient**

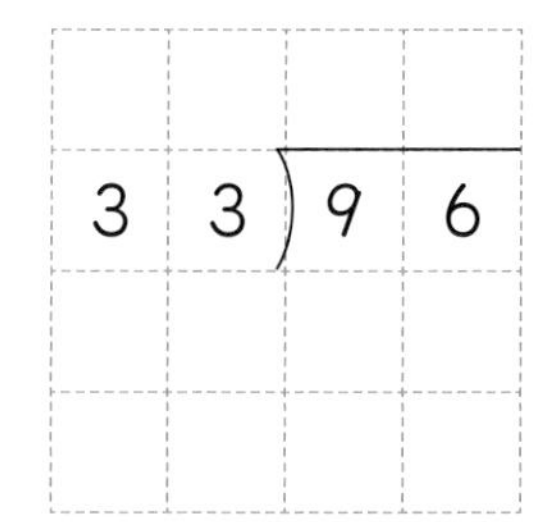

2 Let's think about the division algorithm to calculate $96 \div 33$ in vertical form.

① Let's guess the quotient.

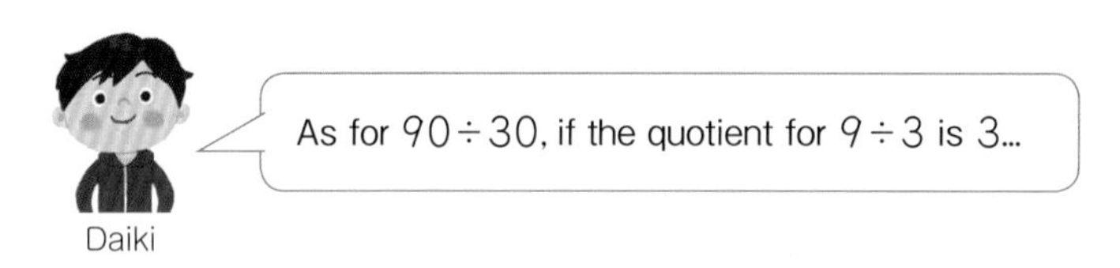

② Let's try the calculation with quotient 3.

$$\begin{array}{r} 3 \\ 33 \overline{)\,96} \\ 99 \end{array}$$

Purpose In vertical form, what should we do if the guess quotient is too large?

$33 \overline{)\,96}$ → $3 \overline{)\,9}$ (tentative quotient 3) → $\begin{array}{r} 3 \\ 33 \overline{)\,96} \\ 99 \end{array}$ Can not subtract. → Make the quotient smaller by 1 unit. → $\begin{array}{r} 2 \\ 33 \overline{)\,96} \\ \underline{66} \\ 30 \end{array}$

30 is smaller than 33.

The first guess of the quotient is called tentative quotient. If the tentative quotient is too large, you have to replace it with a quotient that is smaller by 1 unit.

③ Let's confirm the answer.

Want to confirm

Let's guess the quotient and calculate. Also, let's confirm the answer.

① $56 \div 14$　② $60 \div 12$　③ $68 \div 24$

④ $94 \div 32$　⑤ $67 \div 23$　⑥ $79 \div 13$

Want to think

3 Let's think about the division algorithm to calculate $68 \div 16$ in vertical form.

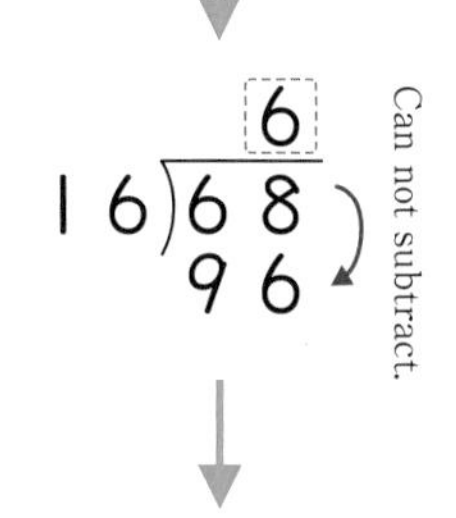

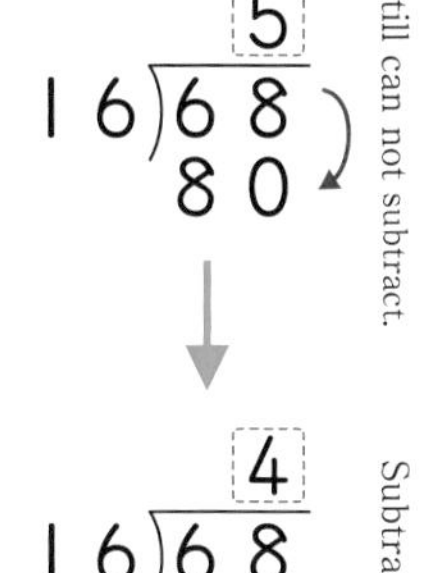

① Let's write a tentative quotient.

② Let's multiply the divisor and the tentative quotient.

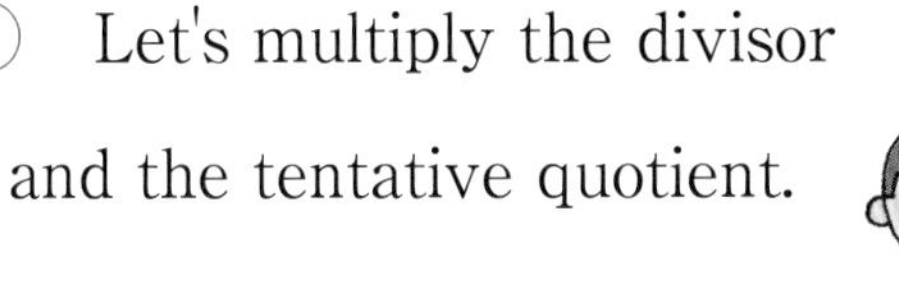

③ Let's replace it with a number that is smaller by 1 unit.

④ Let's make the tentative quotient smaller by 1 unit again.

Summary

If a tentative quotient is too large, replace it with a quotient that is smaller by 1 unit until the correct quotient is found.

Want to confirm

3 Let's guess the quotient and calculate. Also, let's confirm the answer.

① $70 \div 14$　② $69 \div 15$　③ $97 \div 16$

④ $72 \div 15$　⑤ $53 \div 16$　⑥ $82 \div 29$

Want to think The case: (3-digit) ÷ (2-digit)

4 Let's think about the division algorithm to calculate 170 ÷ 34 in vertical form.

$34\overline{)170}$

Purpose Can we divide 3-digit numbers using vertical form in the same way we have learned?

① From which place value can the quotient be written?

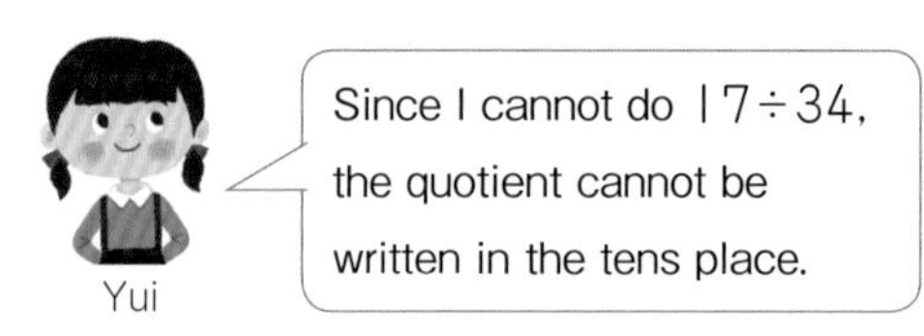

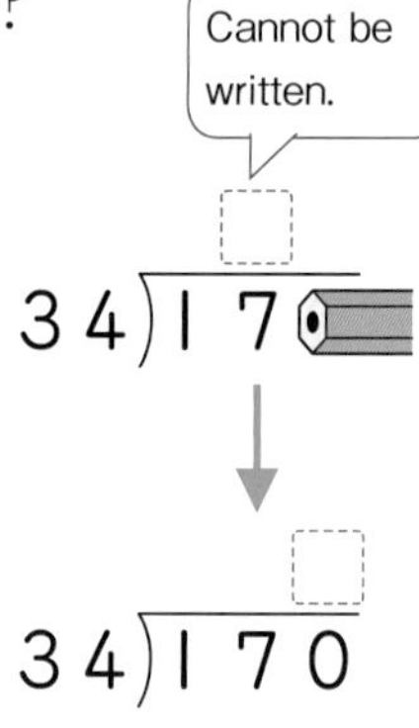

② Let's guess the quotient.

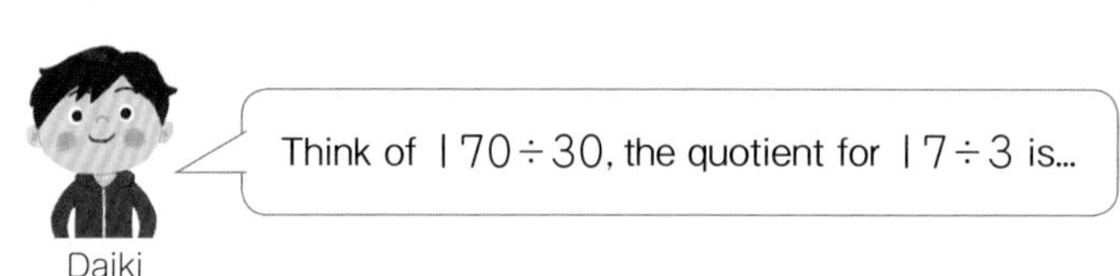

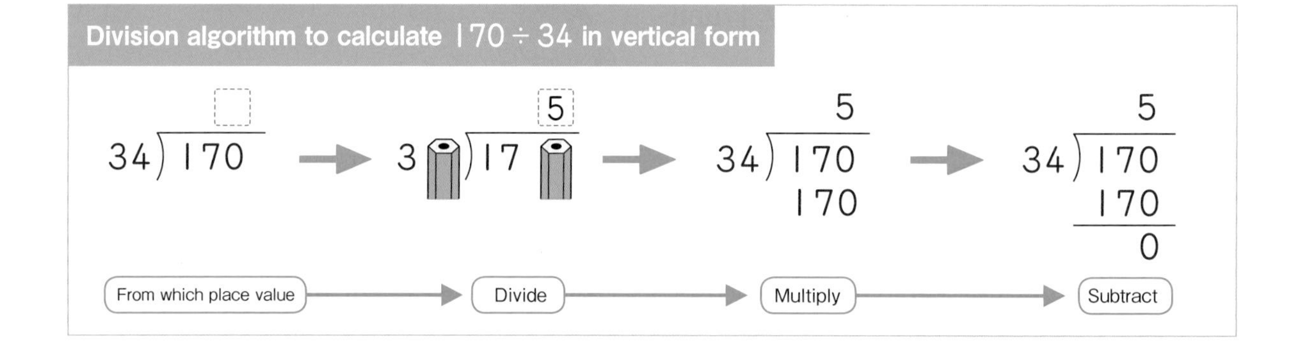

Want to try

Let's think about the division algorithm to calculate 229 ÷ 38 in vertical form.

$38\overline{)229}$

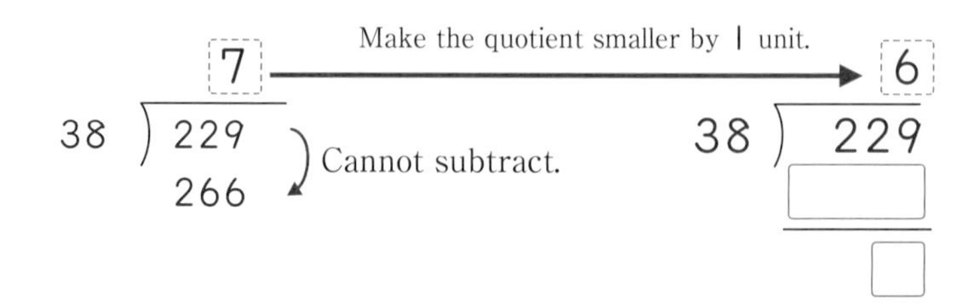

Want to think

5 Let's think about the division algorithm to calculate $326 \div 36$ in vertical form.

① From which place value can the quotient be written?

$36\overline{)326}$

② Let's guess the quotient.

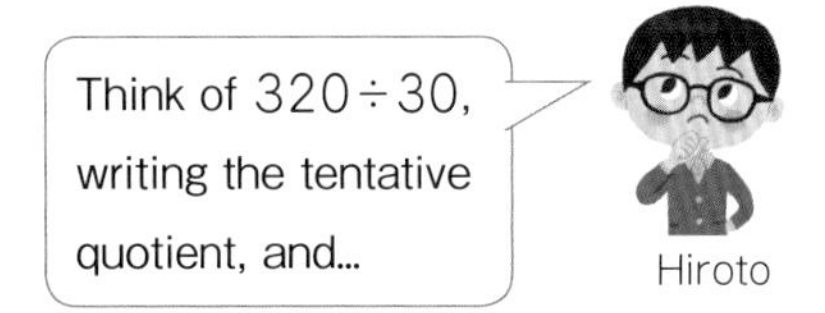

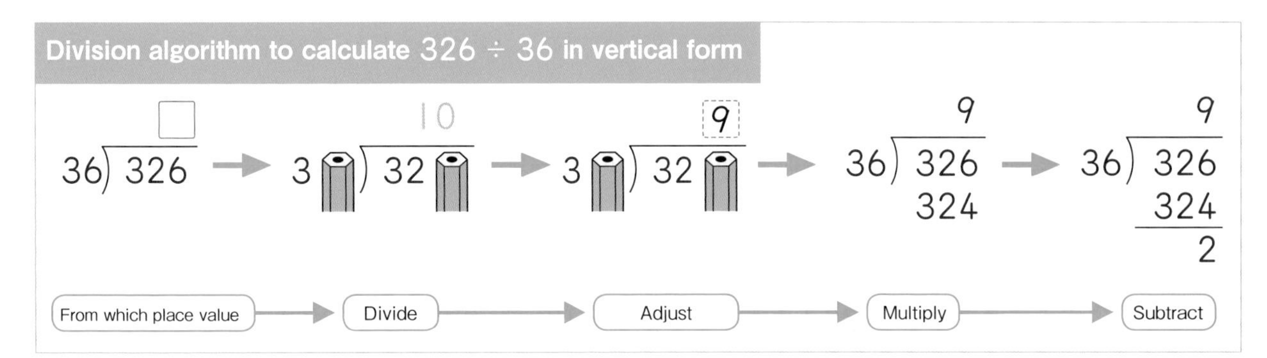

Summary

Even when the dividend has three digits, write the tentative quotient, and find the correct quotient by reducing 1 unit.

Want to confirm

6 Let's solve the following calculations in vertical form.

① $255 \div 51$ ② $284 \div 71$ ③ $201 \div 63$

④ $191 \div 24$ ⑤ $365 \div 48$ ⑥ $208 \div 21$

⑦ $711 \div 78$ ⑧ $217 \div 25$ ⑨ $257 \div 29$

3 Division by 2-digit number (2)

Want to solve Division with a 2-digit quotient

We want to divide 252 sheets of colored paper such that 12 children receive the same number. How many sheets will each child receive?

① Let's write a math expression.

[ ]

② Let's make a guess on how much is the quotient.

③ From which place value can the quotient be written?

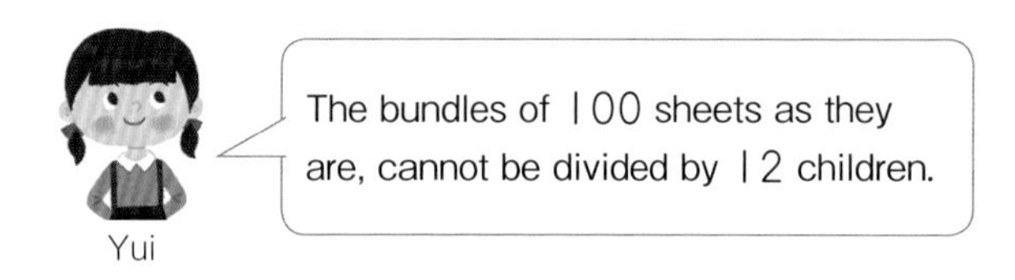

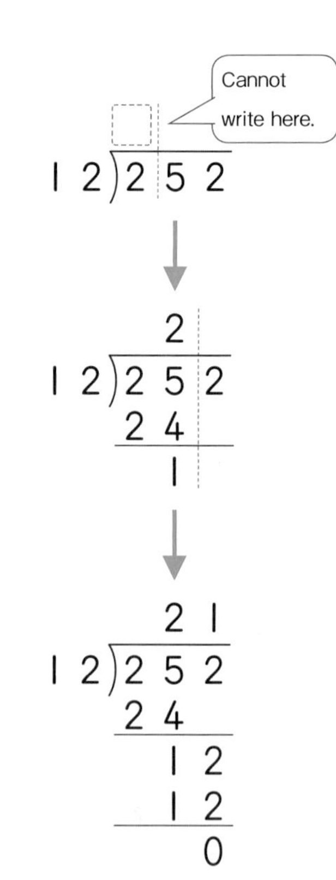

④ Let's change the 100 sheet bundles into 10 sheet bundles and divide by 12 children.

$25 \div 12 =$ [ ] remainder [ ]

⑤ Let's change the remaining 10 sheet bundle into single sheets, add 2 single sheets, and divide by 12 children.

$12 \div 12 =$ [ ]

⑥ How many sheets of colored paper will each child receive?

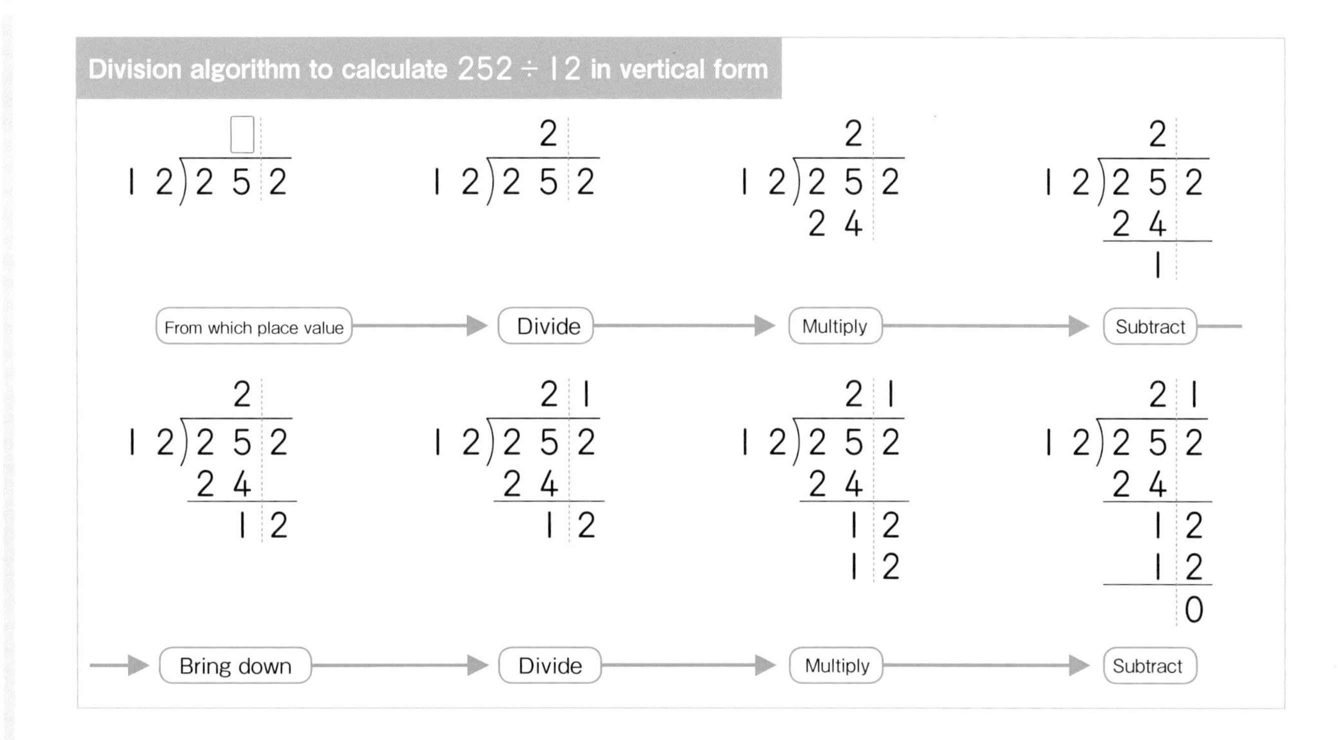

To divide in vertical form, after deciding and writing the quotient, "divide," "multiply," "subtract," "bring down" and repeat the calculation steps.

Want to confirm

1 Let's solve $322 \div 14$ in vertical form.

From which place value can the quotient be written?

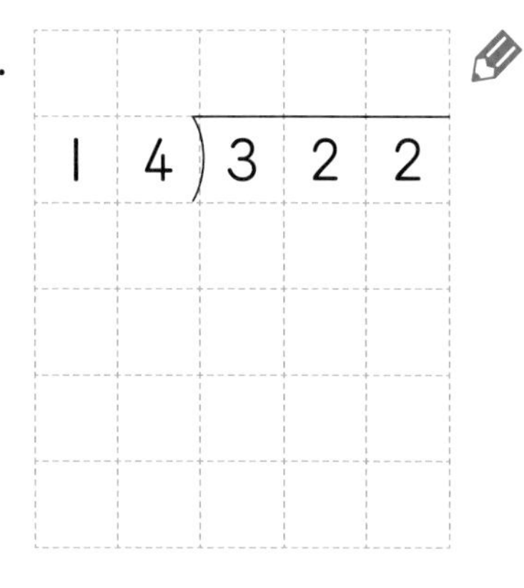

Want to try

2 Let's solve $980 \div 28$ in vertical form.

Want to confirm

3 Let's solve the following calculations in vertical form.

① $736 \div 16$ ② $810 \div 18$ ③ $432 \div 18$

④ $851 \div 26$ ⑤ $798 \div 35$ ⑥ $585 \div 39$

⑦ $612 \div 36$ ⑧ $578 \div 23$ ⑨ $939 \div 37$

Want to know

2 **Let's think about the division algorithm to calculate $607 \div 56$ in vertical form.**

56)60

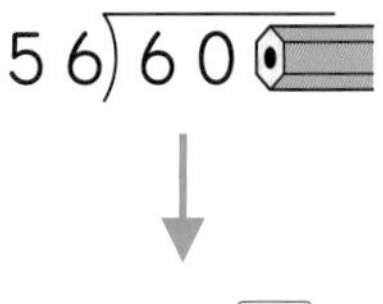

```
    1□
56)607
   56
    47
```

① From which place value is the quotient written?

② What number should be written in the ones place of the quotient?

Want to confirm

4 $859 \div 21$ is being solved in vertical form.

Let's explain ⓐ and ⓑ algorithms shown on the right.

ⓐ
```
     40
21)859
   84
    19
    00
    19
```

ⓑ
```
     40
21)859
   84
    19
```

Want to try

5 Let's solve the following calculations in vertical form.

① $705 \div 34$ ② $913 \div 13$ ③ $856 \div 42$

④ $531 \div 26$ ⑤ $576 \div 56$ ⑥ $942 \div 47$

⑦ $643 \div 16$ ⑧ $750 \div 25$ ⑨ $900 \div 18$

Want to try

Let's correct the mistakes on the following calculations in vertical form.

①

```
        2
22 ) 4 4 6
     4 4
     -----
         6
```

②

```
        2 1
31 ) 6 4 5
     6 2
     -----
       2 5
       3 1
     -----
         6
```

③

```
        1 0
57 ) 7 0 4
     5 7
     -----
       3 4
```

Want to know

Let's think about the division algorithm to calculate 942 ÷ 314 in vertical form.

① From which place value is the quotient written?

② Let's guess the quotient and calculate.

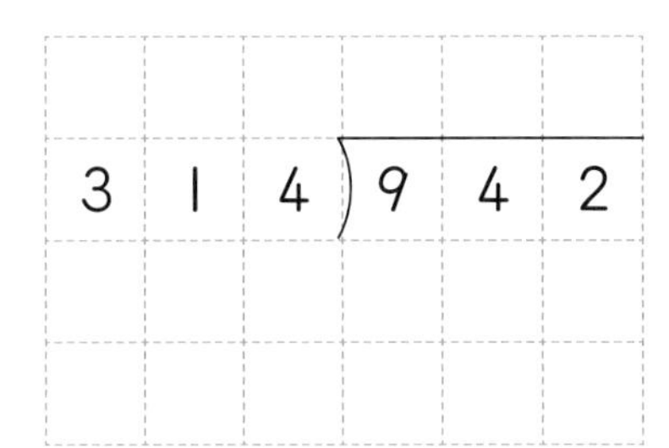

Want to confirm

Let's solve the following calculations in vertical form.

① 744 ÷ 124　② 928 ÷ 232　③ 969 ÷ 323

④ 639 ÷ 213　⑤ 696 ÷ 348　⑥ 952 ÷ 136

⑦ 860 ÷ 213　⑧ 524 ÷ 238　⑨ 950 ÷ 289

Want to know

Activity

1 Let's fill in the □ and ○ with numbers in the following sentence and make math sentences.

There are □ chocolates. If each child receives ○ chocolates, then the chocolates can be distributed to 4 children.

① There are 20 chocolates. Let's write a math sentence when 5 chocolates are given to each child.

$20 \div 5 = 4$

② Let's write various math sentences that become □÷○= 4.

③ Let's look at math sentences with quotient 4, and find a rule.

Purpose As for division sentences with equal quotient, what type of rules are there?

$8 \div 2 = 4$　　$36 \div 9 = 4$

$16 \div 4 = 4$　　$12 \div 3 = 4$

$24 \div 6 = 4$　　$4 \div 1 = 4$

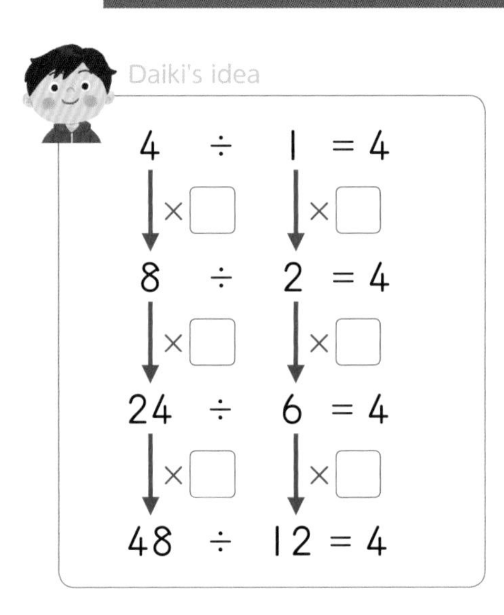

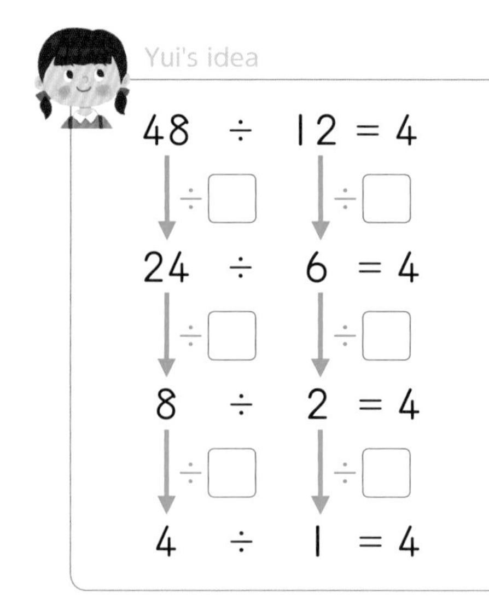

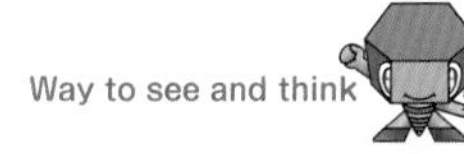

As for a division sentence with equal quotient, find the rule after lining up the dividend and divisor in an increasing and decreasing order.

④ Let's explain the rules found by both children.

Summary

In division, the quotient remains the same if the dividend and divisor are multiplied by the same number. Also, the quotient remains the same if the dividend and divisor are divided by the same number.

Want to confirm

1 Let's use the rules of division to fill in the □ with numbers and find the answer.

① $80 \div 16 = \square \div 8$ ② $90 \div 18 = \square \div 9$

③ $270 \div 90 = 27 \div \square$ ④ $150 \div 25 = \square \div 100$

Want to know

2 **Let's use the rules of division to calculate 24000 ÷ 300 with a new idea.**

$24000 \div 300 = \square$

↓ ÷100 ↓ ÷100

$\square \div \square = \square$

If the dividend and divisor are divided by the same number, then the quotient do not change.

① Let's calculate in vertical form.

```
        8 0
3 0̸ 0̸)2 4 0 0̸ 0̸
       2 4
       -----
           0
```

Way to see and think

I am looking at groups of 100.

As for divisions that include dividends and divisors with 0 at the end, you can only erase the same number of zeros on each and calculate.

Want to confirm

Let's solve the following calculations in vertical form.

① $1500 \div 50$　② $36000 \div 400$　③ $56000 \div 700$

Want to improve

For $3700 \div 500$, let's use the rules of division and calculate with a new idea.

① Let's think about how to calculate.

```
          7
5 0̸ 0̸)3 7 0̸ 0̸
       3 5
       ----
         2
```

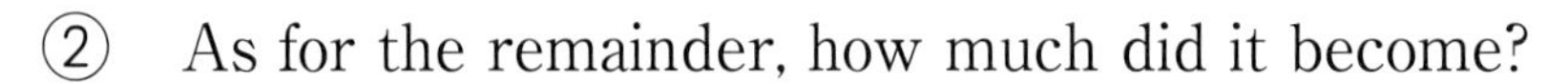

② As for the remainder, how much did it become?

$3700 \div 500 = \square$ remainder $\square$

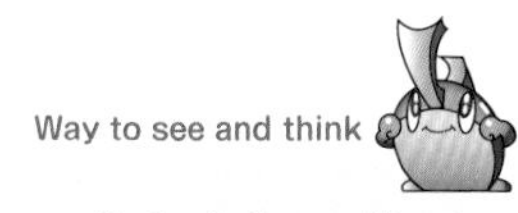

Calculating with
100 as one unit.

③ Let's confirm the answer.

$500 \times \square + \square = \square$

Divisor　Quotient　Remainder　Dividend

As for divisions in which 0 are erased, when finding the remainder, only place the erased number of zeros into the remainder.

Want to confirm

For $6200 \div 400$, let's use the rules of division and calculate with a new idea. Also, let's confirm the answer.

Want to try

Let's solve the following calculations in vertical form.

① $1700 \div 800$　② $4400 \div 600$　③ $7500 \div 80$

Division in various countries

The 2 calculations below show division methods in other countries. Let's compare with Japan's division algorithm in vertical form.

$984 \div 23$

① Let's explain the division methods in Germany and Canada.

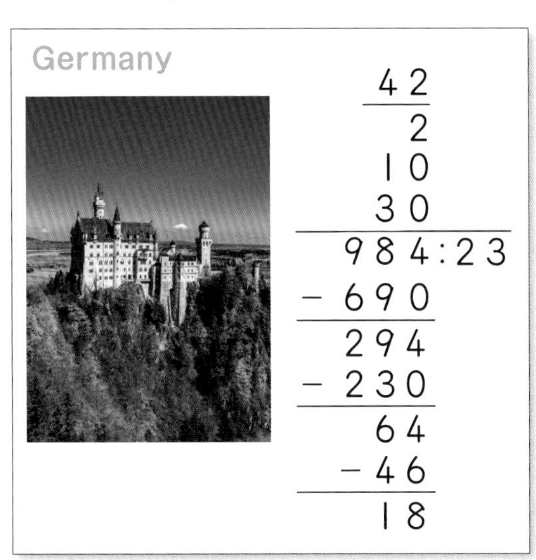

In Germany, they start with a small tentative quotient and then repeat.

② Compare the division methods in Japan, Germany, and Canada.

Let's discuss their good points.

Doing division in Germany seems to take more time but with less mistakes.

23)984

③ Let's try to calculate $898 \div 28$ using the method in each country.

What you can do now

Can solve a division with a 2-digit divisor.

1 Let's solve the following calculations in vertical form.

① $60 \div 20$　② $140 \div 70$　③ $130 \div 40$

④ $96 \div 32$　⑤ $97 \div 27$　⑥ $85 \div 19$

⑦ $344 \div 43$　⑧ $385 \div 56$　⑨ $411 \div 45$

⑩ $672 \div 28$　⑪ $453 \div 17$　⑫ $738 \div 24$

Can make a division expression and find the answer.

2 Let's answer the following questions.

① If you buy some bread, each piece will cost 75 yen. If the total amount payed is 900 yen, how many pieces of bread will you buy?

② 342 screws were placed inside bags in groups of 18 screws. How many bags were needed?

③ 800 stickers were divided into 12 children. How many stickers did each child receive and what was the remainder?

Understanding the rules of division.

3 Using the rules of division, let's fill in the □ with numbers.

① $18 \div 2 = 9$

↓ × □　↓ × 3

$54 \div 6 =$ □

②

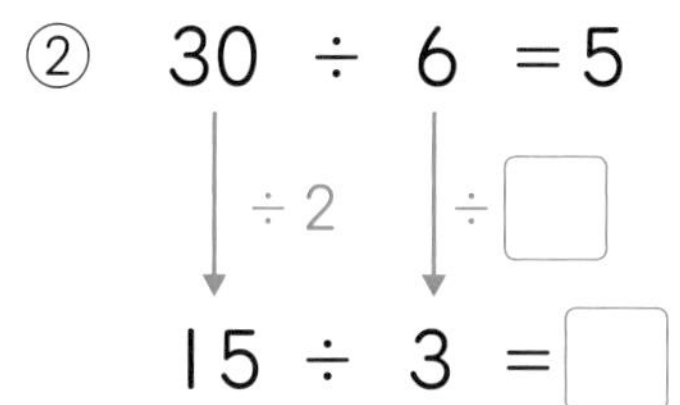

③ $12 \div 3 = 24 \div$ □

④ $18 \div 6 =$ □ $\div 2$

⑤ $80000 \div 200 =$ □ $\div 2$

⑥ $3200 \div 160 = 320 \div$ □

Supplementary problems p. 155

Usefulness and efficiency of learning

1 Let's solve the following calculations in vertical form.

① $64 \div 21$ ② $74 \div 15$ ③ $505 \div 55$

④ $715 \div 42$ ⑤ $567 \div 28$ ⑥ $736 \div 36$

☐ Can solve a division with a 2-digit divisor.

2 We want to divide 113 eggs such that 12 children receive the same number. How many eggs will each child receive and what is the remainder?

☐ Can make a division expression and find the answer.

3 From a tape with 7 m 60 cm length, how many tapes with a 50 cm length can you get? How many cm is the remainder?

☐ Can make a division expression and find the answer.

4 Kaori's class has 38 children. When 570 sheets of origami are divided, how many sheets of origami will each child receive? Let's also answer if there is a remainder.

☐ Can make a division expression and find the answer.

5 Let's explain the reasons why $320 \div 40$ can be calculated using $32 \div 4$.

☐ Understanding the rules of division.

6 On the following table, the vertical, horizontal, and diagonal products have the same answer.

Let's fill in each empty space with a number.

☐ Can solve a division with a 2-digit divisor.

12	ⓐ	2
ⓑ	6	36
18	ⓒ	ⓓ

How many times: Distance of Jump

Want to compare

1 **When Yuma did a long jump, the distance of his jump was 260 cm. Yuma's height is 130 cm. Let's compare his jump distance and height.**

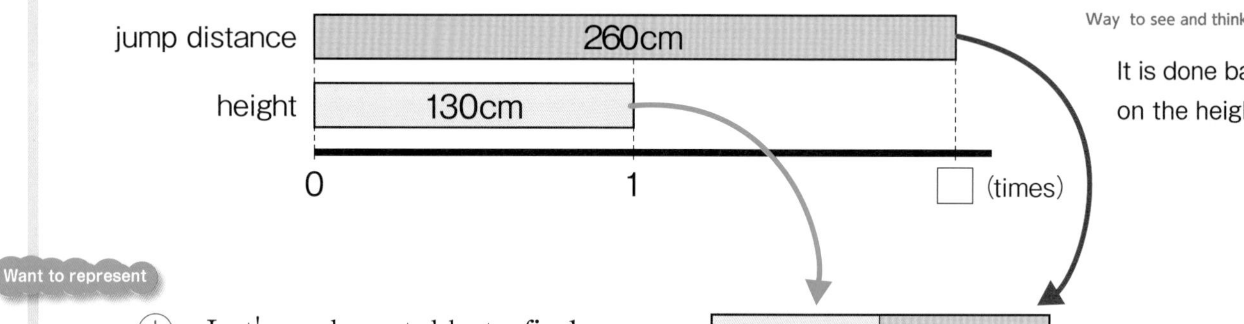

Way to see and think

It is done based on the height.

Want to represent

① Let's make a table to find the relationship between the distance of the jump and height.

130cm	260cm
1 time	□ times

Want to think

② Using ①, let's think a way to find how many times of his height he jumped.

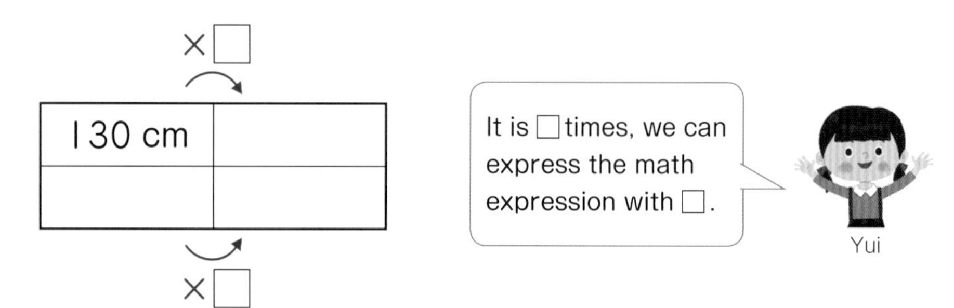

③ How many times of his height did he jump?

Want to confirm

An athlete jumped 8 m 50 cm in a long jump competition.

The athlete's height is 170 cm.

Let's compare this athlete's height and distance of the jump.

① Let's make a table to find the relationship between the distance of the jump and height.

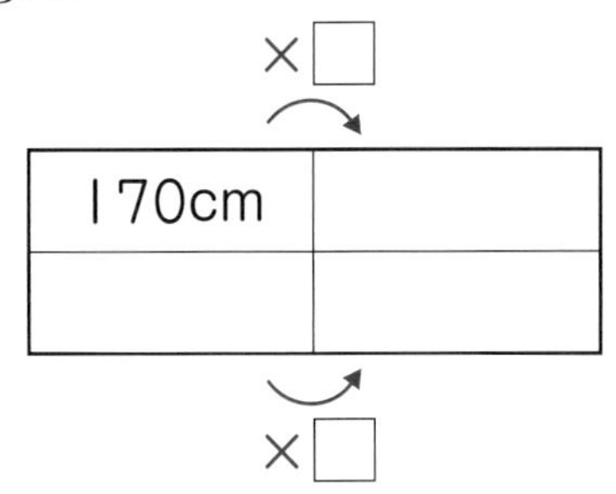

170cm	

② Let's write a math expression to find how many times of her height she jumped.

③ How many times of her height did she jump?

Want to try

A frog can jump 40 times its body length.

Let's think about the statement below.

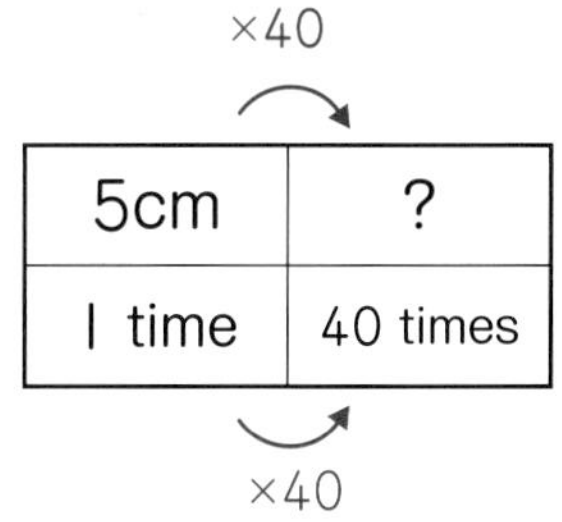

5cm	?
1 time	40 times

① If the body length of the frog is 5 cm, then how many meters can it jump?

② Also, if you could jump the same as the frog, 40 times your own height, how many m and cm would that be?

Let's interpret information correctly from a line graph.

Want to explore School's Library

The library committees from School A and School B are promoting cooperative reading activities. The following data is a summary of the numbers of books borrowed in a 4 month period from April to July.

Numbers of books borrowed in both schools in the 4 month period from April to July.

Table 1 Numbers of books borrowed in both schools by month (books)

School \ Month	April	May	June	July	Total
School A	986	2918	3414	2420	9738
School B	849	2523	2938	2095	8405

Table 2 Numbers of books borrowed in School A by type (books)

Story	Science	History	Biography	Other	Total
3800	1977	1496	989	1476	9738

Numbers of "Story" books borrowed in the 4 month period from April to July

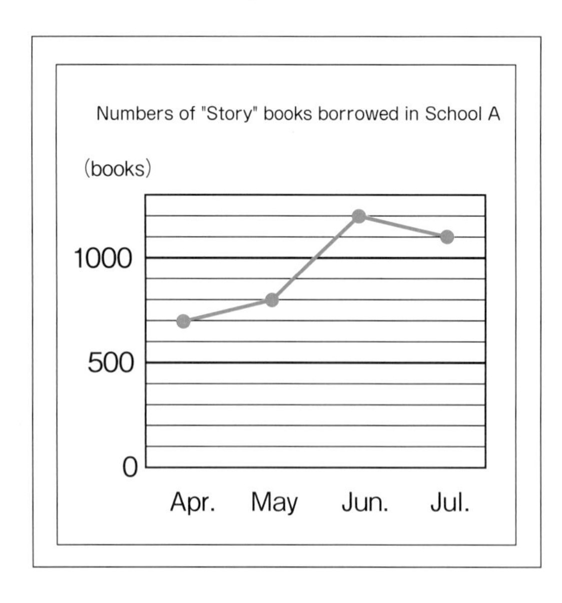

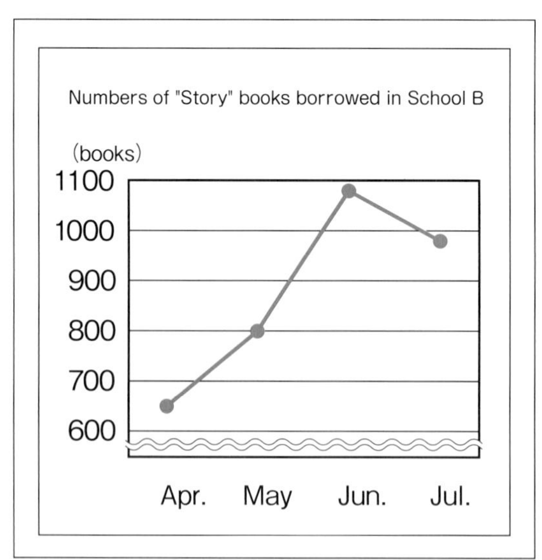

Daiki said the following based on a comparison done from May to June on the numbers of "Story" books in both graphs.

Compared to School A in the period from May to June, School B has a sudden increase. Therefore, compared to School A in the period from May to June, School B's number of books borrowed increased more.

Daiki

Think by myself

1 Is Daiki's opinion correct? Based on your interpretation from the graph, let's write down your reason using numbers and words.

Think in groups

2 Make a group and choose one from A, B, and C, and let's discuss.

Ⓐ Examine how the graph inclines on page 27 of this book and let's check whether Daiki's opinion was correct.

Ⓑ In School A, from May to June, how much did the number of "Story" books borrowed increase? Let's interpret from the graph.

Ⓒ In School B, from May to June, how much did the number of "Story" books borrowed increase? Let's interpret from the graph.

Think as a class

3 From the discussion explained in groups in exercise 2, let's think about whether Daiki's opinion was correct.

Then let's discuss what you need to be careful about to compare the inclination manner of graphs.

Divided in groups or pairs, if you share, various things can be investigated in a short time.

01403

Exactly 20000 shots?

As for approximate numbers, in how much range can we use?

8 Round Numbers

Let's think about how to represent and calculate approximate numbers.

1 How to represent round numbers

Want to represent

1 **As for the numbers of elementary school students in East and West Town, how many thousand students are there approximately in each town?**

West Town	2368 people
East Town	2825 people

Both are numbers between 2000 and 3000.

From which is closest? 2000 or 3000?

Purpose What is the best way to represent an approximate number for a number of people?

① Let's think based on the following number line.

2000 3000

An approximate number is called a **round number**. When you need to know a rough number, you can represent it by round numbers. Approximately 2000 is said to be about 2000.

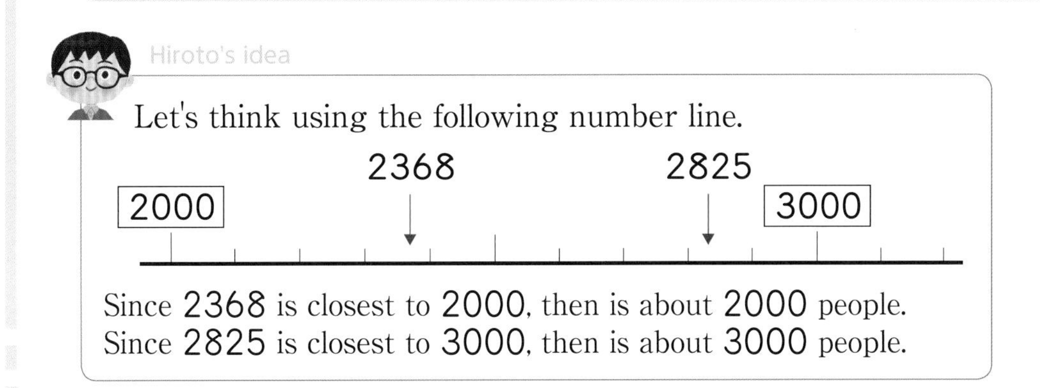

Want to explain

② To find if the number is closer to either 2000 or 3000, what place value number should we look at?

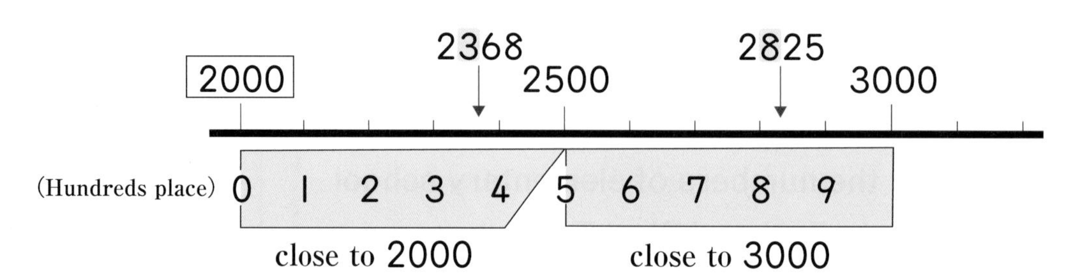

To represent the approximate number of thousands by round numbers, it is called rounding to the nearest thousand.

How to represent round numbers

When we want to represent a round number to the nearest thousand, we have to look at the number in the hundreds place.

When the number in the hundreds place is less than 5 (0, 1, 2, 3, or 4) we can leave the number in the thousands place unchanged and replace the numbers to the right with 000.

When the number in the hundreds place is greater than or equal to 5 (5, 6, 7, 8, or 9), we add 1 unit to the number in the thousands place and replace the numbers to the right with 000.

The way of representing round numbers shown above is called **rounding off**.

Greater than or equal to 5 means larger than 5 or equal to 5.

Less than 5 means smaller than 5, not including 5.

Less than or equal to 5 means smaller than 5 or equal to 5.

Summary

For numbers between 2000 and 3000, its representation by round numbers to the nearest thousand when the hundreds place is: less than 5, it can be rounded off to about 2000, greater than or equal to 5, it can be rounded off to about 3000.

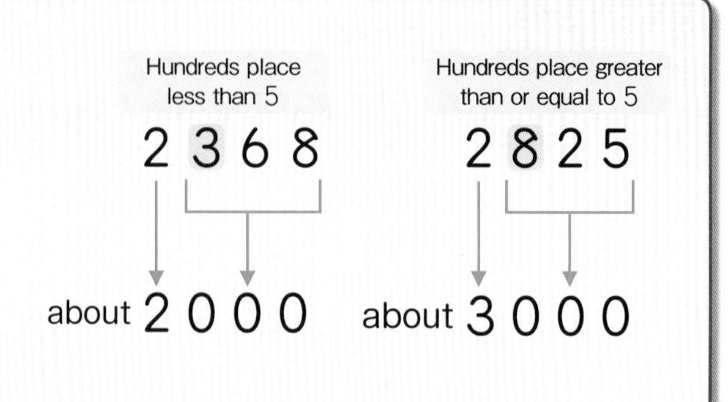

Want to confirm

Let's round off to the nearest thousand round number.

① 3472 ② 6728 ③ 4856 ④ 8543 ⑤ 973

Want to represent

2 **The table on the right represents the number of elementary school students in Aichi and Aomori Prefectures in 2017. About how many ten thousand students are there in each prefecture? Let's round off to the nearest ten thousand round number.**

Aichi Prefecture	414767 students
Aomori Prefecture	59233 students

Want to confirm

Let's round off to the nearest ten thousand round number.

① 68492 ② 53862 ③ 872691 ④ 236389

Want to represent Round number from the highest first digit

Let's round off from the highest first digit and highest two digits round number. Let's think how many digits should be rounded off from the highest place value. Let's write on the following table.

From the highest first digit
7 869

From the highest two digits,
78 69

	7869	4139	52630
From the highest first digit round number	8000		
From the highest two digits round number	7900		

Want to confirm

Let's round off the following numbers from the highest first digit and highest two digits round number.

① 2642 ② 8468 ③ 35418 ④ 62843

Want to know Range of a represented round number

3 When rounded off to the nearest hundred round number, let's explore whole numbers that become 800.

① When rounded off to the nearest hundred round number, from all the whole numbers that become 800, which is the largest and smallest number?

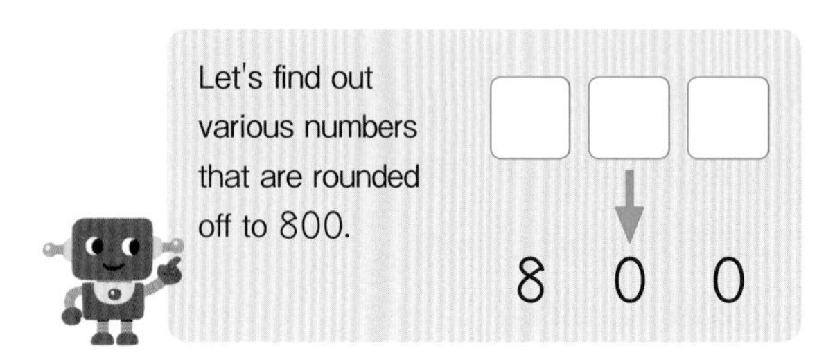

If the number in the tens place is greater than or equal to 5, then the number in the hundreds place will increase one unit.

When rounded off to the nearest hundred round number, the whole numbers that become 800 are from 750 to 849. This number range is called "greater than or equal to 750 and less than 850."

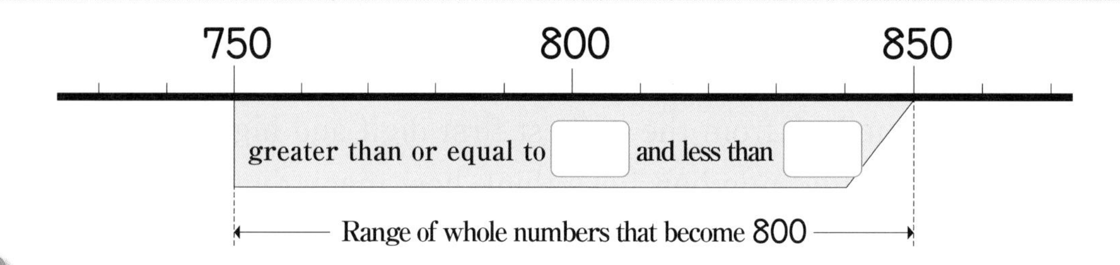

Want to confirm

Let's fill in the ☐ with numbers.

When rounded off to the nearest thousand round number, the numbers that become 34000 are the numbers greater than or equal to ☐ and less than ☐.

Words

More than, within, after
until here.
Includes the point itself.

Less than, future, unknown
not yet.
Do not include the point itself.

2 Rounding down and rounding up

Want to represent

1 **There are 876 sheets of paper, which are to be bundled every 100 sheets. How many sheets of paper can be bundled?**

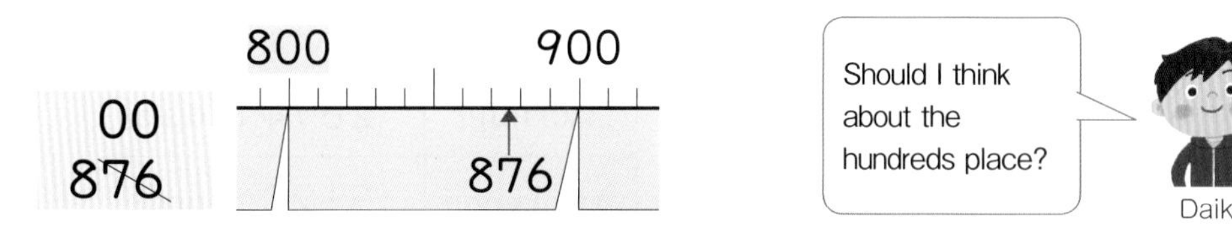

The numbers that do not reach 100 are lowered to 0. This is called "**rounding down** to the nearest hundred round number."

Want to try

1 There were 823 people who went on a trip by train. One train car could carry 100 people. How many cars were used?

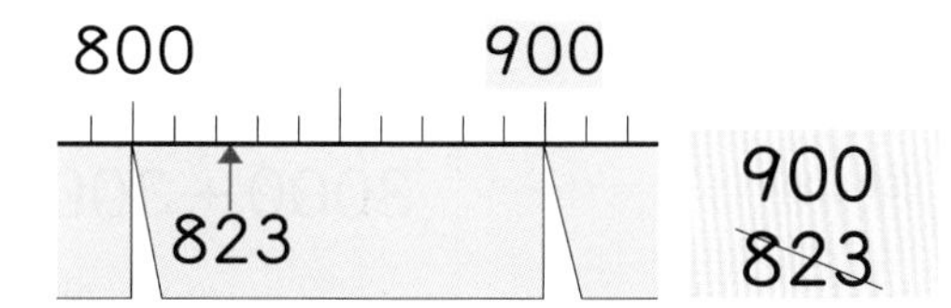

The numbers that do not reach 100 are increased to 100. This is called "**rounding up** to the nearest hundred round number."

As shown above, the way of rounding numbers is not limited to only rounding off, but also **rounding down** and **rounding up**.

Want to confirm

2 Let's round down the following numbers to the highest two digits round number. Also, let's round up to the highest first digit round number.

① 28138　② 3699　③ 42500　④ 9810

3 Rough estimate

Want to know Adding and subtracting rough estimate

Activity

1

The table below shows the number of visitors in a zoo for one day.

Let's find the total number and the difference by using round numbers.

Number of visitors in a zoo

Morning	2894 people
Afternoon	3128 people

① About how many thousand people visited the zoo that day?

Hiroto's idea

I added the number of people in the morning and afternoon,

$$2894+3128$$
$$=6022$$

and rounded off the sum to the nearest thousand round number and got about 6000 people.

Yui's idea

I rounded off the number of people in the morning and afternoon,

2894 → 3000
3128 → 3000

and added both numbers,

$3000+3000=6000$ (people)

Therefore, about 6000 people.

② About how many hundred more people visited in the afternoon than in the morning?

When you want to find the sum and the difference by using round numbers, first round each number and then calculate.

A number calculated by using round numbers is called a **rough estimate.**

Want to think Multiplication estimate 

2 **There are 315 children going on a field trip. The train ticket costs 190 yen for each child. How much will the cost be for all the children?**

① Let's write a math expression.

② Let's use round numbers and apply on the idea of the product.

The idea applied is called "**estimate.**"

Round the multiplier and the multiplicand to the highest first digit round number to estimate the product.

$190 \times 315 \rightarrow 200 \times 300$

③ Let's compute 190×315 using a calculator and compare the answer with the estimate.

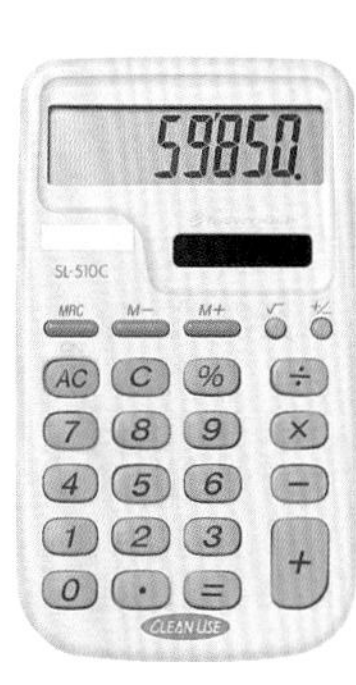

      5 

Want to confirm

 Let's estimate the following products.

① 498×706 ② 2130×587

Want to think Division estimate

At the zoo, the weight of an African elephant is 6270 kg. The weight of Yoshiko is 38 kg.

How many times Yoshiko's weight is the weight of the elephant?

① Let's make a math expression. [ ]

② Let's estimate the quotient by rounding the dividend and the divisor to the highest first digit round number.

$$6270 \div 38 \rightarrow 6000 \div 40$$

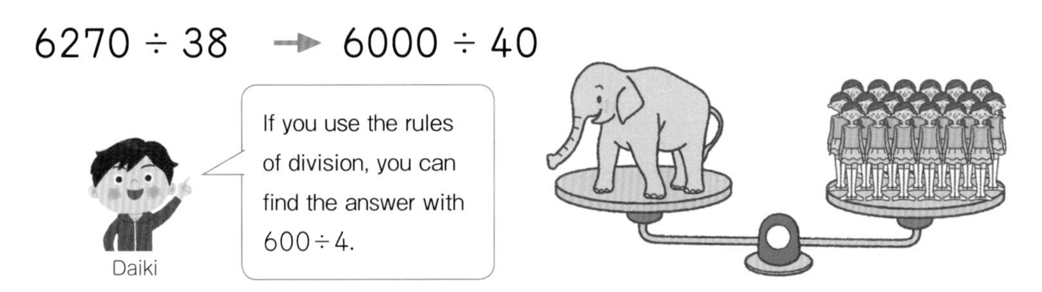

③ Let's compute 6270 ÷ 38 using a calculator, and compare the answer with the estimate.

Want to confirm

About how many times the height of the Statue of Liberty in New York City is the height of the Tokyo Skytree?

634m

46m

Want to try

Let's estimate the following quotients.

① 37960 ÷ 78　　② 90135 ÷ 892

Want to know Using estimates

3 **A family will visit the zoo together. Their anticipated expenses are shown on the right. About how much money should they bring? Let's estimate the amount of money.**

Expenses

Item	Amount (yen)
Train ticket	2960
Entrance fee	2350
Meals	3800

Which method for rounding numbers shall we use so that they have enough money?

Want to confirm

5 The family went shopping before going to the zoo. If they spend more than 1000 yen in the store, they will receive one free pass to the zoo. The table on the right shows the shopping list. Can they receive a free pass?

Shopping List

Item	Amount (yen)
Chocolate	154
Potato Chips	148
Animal Book	660

Which method of rounding shall we use to know whether they will spend more than 1000 yen?

Want to explore

1 **The summarized results in the table below show the numbers of elementary and junior high school students in Numazu City. Let's represent the data with a graph.**

① How many students should each scale represent? Let's write round numbers on the right column.

② Let's draw a line graph.

Numbers of elementary and junior high school students in Numazu City

Year	Numbers of elementary and junior high school students (people)	Round number (people)
1994	21643	
1996	20566	
1998	19430	
2000	18531	
2002	17771	
2004	17135	
2006	17176	
2008	16838	
2010	16150	
2012	15473	
2014	14757	
2016	14110	

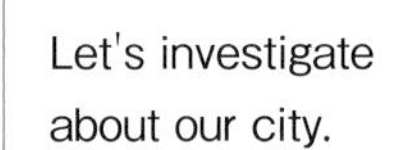

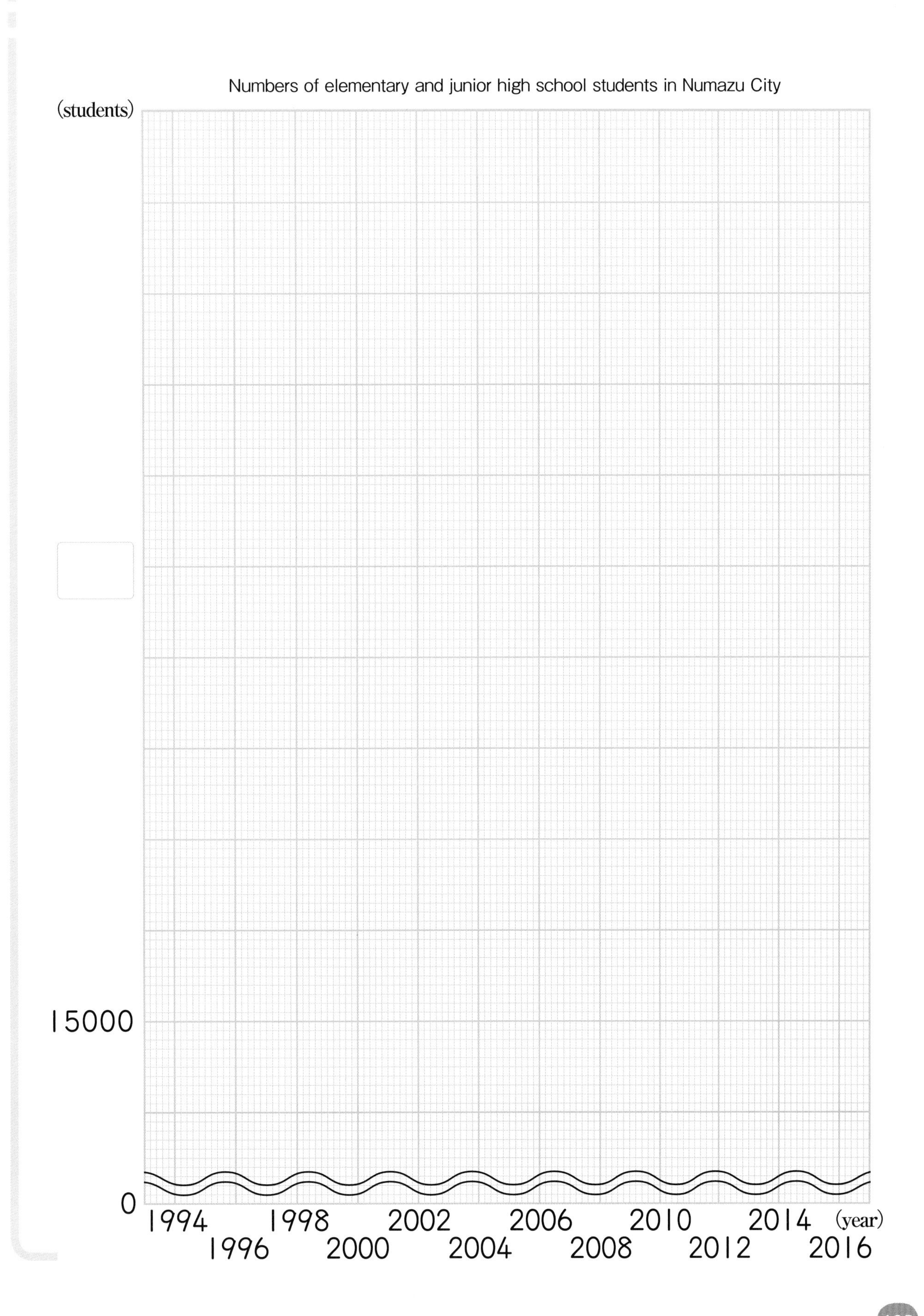

Numbers of elementary and junior high school students in Numazu City
(students)
15000
0
1994
1996
1998
2000
2002
2004
2006
2008
2010
2012
2014
2016
(year)

What you can do now

☐ **Understanding how to represent round numbers.**

1 Let's round the following numbers.

① Let's round off to the nearest ten thousand round number.

ⓐ 47560 ⓑ 623845 ⓒ 284999

② Let's round off to the nearest thousand round number.

ⓐ 36800 ⓑ 513291 ⓒ 49781

③ Let's round off to the highest two digits round number.

ⓐ 67325 ⓑ 748500 ⓒ 195000

④ Let's round down to the nearest thousand round number.
Also, to the nearest ten thousand round number.

ⓐ 36420 ⓑ 43759 ⓒ 239500

⑤ Let's round up to the highest first digit round number.
Also, to the highest two digits round number.

ⓐ 4586 ⓑ 62175 ⓒ 832760

☐ **Can solve problems using round numbers.**

2 There are 789 thousand-yen bills, which are to be bundled every 10 bills. How many bills can be bundled? Also, how many yen is the sum of the bundles?

Supplementary problems ➡ p.156

Words

【概数】 This kanji means "round number." The kanji "概" means "approximately" or "about."

【約】 This kanji means "more or less" or "about."

Usefulness and efficiency of learning

1 Is the way of using round numbers correct in the following statements? If correct, let's place a ○.

☐ Understanding how to represent round numbers.

① ( ) Since my arithmetic exam score is 68 points, I can say it is about 100 points.

② ( ) Since the number of books in my school's library is 8725 books, I can say there are about 9000 books.

2 Let's answer the questions about the following numbers.

☐ Understanding how to represent round numbers.

38478, 37400, 38573, 37501,
38500, 37573, 38490, 37499

① Which numbers become 38000 when rounded off to the nearest thousand round number?

② Which numbers become 37000 when rounded down to the nearest thousand round number?

③ Which numbers become 39000 when rounded up to the nearest thousand round number?

3 Let's answer the following questions.

☐ Can solve problems using round numbers.

① I think to buy one 1980 yen shirt, one 893 yen pair of socks, and one 3480 yen pair of pants.
How many thousand yen is it enough to bring?

② The visitors to the Art Museum were 1456 people on Saturday and 2438 people on Sunday. On which day were the visitors more and by how many? Let's answer by rounding off to the nearest hundred round number.

③ When you measured the length around a tree, it was 147 cm. When you measured the length around another tree, it was 28 cm. About how many times thicker is the thicker tree compared to the thinner tree? Let's round off each length to the nearest ten and find the anwer.

Division in vertical form: thinking with round numbers

Yui used round numbers to calculate $84 \div 39$ in vertical form. Her idea is shown on the right.

Let's calculate $84 \div 39$ using Yui's tentative quotient.

Rounded off 84 and 39, and made a tentative quotient with $80 \div 40$.

$80 \div 40 = 2$

Therefore, I think the tentative quotient is 2.

Hiroto used round numbers to calculate $68 \div 16$ in vertical form. His idea is shown below.

Hiroto's idea.

Rounded off 68 and 16, and made a tentative quotient with $70 \div 20$.

$70 \div 20 = 3$ remainder 10

Therefore, I think the tentative quotient is 3.

$$16 \overline{)68} \quad \text{quotient } 3$$

① Let's continue Hiroto's vertical form process.

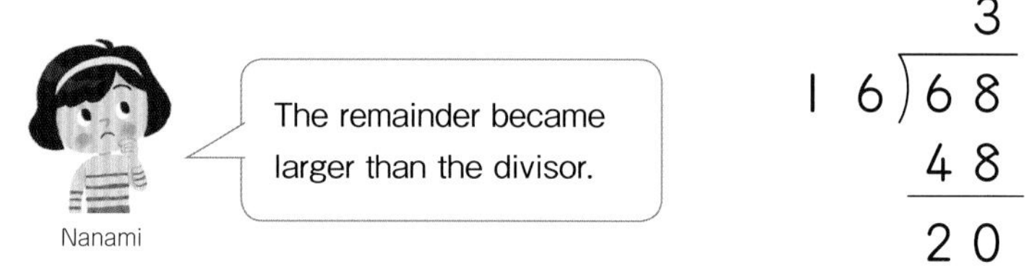

$16\overline{)68}$ → quotient 3: $16\overline{)68}$ → quotient 3: $16\overline{)68}$, 48, 20 → quotient 4: $16\overline{)68}$, 64, 4

Make the quotient larger by 1 unit.

20 is larger than 16.

② Based on exercise 3 from page 107, let's discuss the differences between the way with pencils and Hiroto's.

Which injury happens the most?

To summarize and understand better, what shoud we do?

9 Arrangement of Data

Let's think about how to summarize tables.

1 Arrangement of table

Want to explore

1 **Let's organize the data in the table on the right and explore the situation of injuries in the school.**

Want to summarize

① Let's explore the place of the injuries.

ⓐ Let's write down on the table below where the injuries happened most frequently.

ⓑ Let's look at the table and explain what you have noticed.

Number of children and place of injury

Place of injury	Number of children (people)
Playground	一
Corridor	T
Classroom	
Gymnasium	
Stairs	
Total	

Write the character "正" to organize as you did in the 3rd grade.

Daiki

Record of injured children

Grade	Time	Place	Kind of injury
5	Morning	Corridor	Bruise
4	Break time	Playground	Scratch
5	Break time	Corridor	Bruise
1	Noon break	Classroom	Scratch
3	During class	Playground	Scratch
3	During class	Playground	Bruise
6	During class	Gymnasium	Scratch
5	Morning	Classroom	Cut
4	Break time	Playground	Scratch
5	During class	Gymnasium	Scratch
3	During class	Gymnasium	Bruise
1	During class	Classroom	Cut
2	During class	Playground	Scratch
6	During class	Gymnasium	Sprain
6	After class	Playground	Sprained finger
5	Morning	Classroom	Cut
5	Break time	Classroom	Scratch
3	Break time	Stairs	Bruise
4	During class	Gymnasium	Sprain
2	During class	Playground	Bruise
6	During class	Gymnasium	Scratch
4	Noon break	Corridor	Bruise
4	Morning	Stairs	Scratch
2	During class	Gymnasium	Sprained finger
1	During class	Playground	Sprain
6	Noon break	Corridor	Cut
5	Break time	Classroom	Scratch
3	During class	Playground	Bruise

I want to make use of it

② Let's explore the type of injury.

ⓐ Let's write down on the table which injuries happened most frequently.

ⓑ Let's look at the table and explain what you have noticed.

Way to see and think

It seems easier to see both perspectives with a summary.

Number of children and kind of injury

Kind of injury	Number of children (people)
Cut	
Bruise	
Scratch	
Sprained finger	
Sprain	
Total	

Want to know

2 **Let's explore the kind of injury and where it happened.**

Keep an eye on both the place and kind, let's fill in the table with a number for the places and kinds of injuries.

Places and kinds of injuries (people)

Place \ Kind	Cut	Bruise	Scratch	Sprained finger	Sprain	Total
Playground						
Corridor						
Classroom						
Gymnasium						
Stairs						
Total						

① Where and which kind of injury has the largest number?

② Where did the most frequent kind of injury happen?

③ Compared to two tables summarized in **1**, what did you find in this table?

2 Arrangement of data

Want to know

1 **Masao asked his classmates to draw a ○ on the card to investigate if they have goldfish (G) or small birds (SB) at home. Let's answer the following questions.**

① What kind of groups can be made depending on the way the ○ are marked.

Ⓐ Who drew two ○ and how many children are there in this group?

Ⓑ Who drew one ○ and how many children are there in this group?

Ⓒ Divide the children who drew one ○ into those who have goldfish and those who have small birds. How many children are there in each group?

Ⓓ Who did not draw any ○ and how many children are there in this group?

② Let's fill in the following table with the corresponding number of children.

Investigation of owned animals (children)

		Goldfish: Having	Goldfish: Not having	Total
Small birds	Having	2		
	Not having			
Total				

③ How many children only have birds?

④ How many children only have goldfish?

What you can do now

☐ **Can make a summary to understand two things.**

1 The following table shows the record of injuries for 4th grade in Soma's school. Let's summarize the data in the table below.

Record of injured children

Name	Place	Kind of injury	Name	Place	Kind of injury
Maeda	Playground	Scratch	Oishi	Playground	Bruise
Ota	Classroom	Cut	Yamada	Playground	Cut
Oba	Classroom	Scratch	Yamamoto	Gymnasium	Scratch
Tanaka	Gymnasium	Sprain	Morita	Gymnasium	Bruise
Nakamura	Corridor	Bruise	Kobayashi	Classroom	Scratch
Noguchi	Gymnasium	Sprained finger	Hayashi	Gymnasium	Scratch

Places and kinds of injuries (Children)

Place \ Kind	Scratch					Total
Playground						
Total						

☐ **Can read a summary from a table.**

2 Mitsuki investigated about the brothers and sisters of her classmates. There are 36 children in her class.

Children with older brothers: 12 children

Children with older sisters: 9 children

Children without older siblings: 18 children

Complete the table on the right with numbers of children.

(Children) 

		Older brother: Yes	Older brother: No	Total
Older sister	Yes			
	No			
Total				36

Supplementary problems p.157

Usefulness and efficiency of learning

1 The following record shows the lunch food and fruit chosen and ordered for the Children Association Fun Party.

Fun Party orders

Food	Fruit	Food	Fruit	Food	Fruit
Rice ball	Apple	Hot dog	Strawberry	Hot dog	Orange
Rice ball	Orange	Rice ball	Banana	Rice ball	Orange
Sandwich	Strawberry	Sandwich	Strawberry	Hot dog	Strawberry
Hot dog	Strawberry	Hot dog	Orange	Hot dog	Banana
Sandwich	Banana	Hot dog	Strawberry	Sandwich	Apple
Sushi roll	Orange	Sushi roll	Apple	Rice ball	Orange
Rice ball	Orange	Hot dog	Apple	Rice ball	Orange

① Let's summarize in the following table.

☐ Can make a summary to understand two things.

Fun Party orders (Children)

Food \ Fruit	Apple	Orange	Strawberry	Banana	Total
Rice ball					
Sandwich					
Hot dog					
Sushi roll					
Total					

② From the foods chosen by the children, which is the one with the highest number? Also, from the fruits chosen by the children, which is the one with the highest number?

☐ Can read a summary from a table.

③ Let's discuss what you understood from the table.

Let's deepen.

Can I use a summary table with information from our surroundings?

Number of passengers

Want to know

Below, 4 children are talking about the number of passengers that ride a train. Let's think about the numbers of passengers.

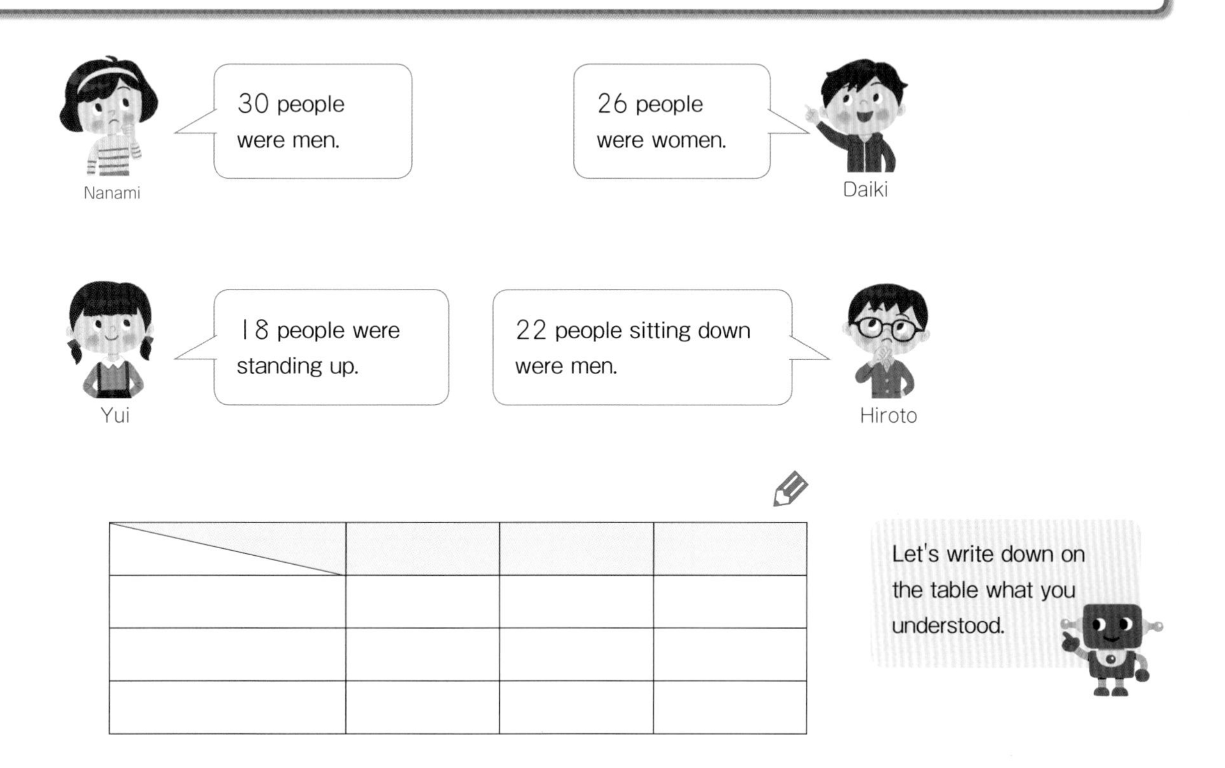

① How do you think in order to find the total number of passengers on this train? Let's explain your answer.

② How do you think in order to find the number of standing men? Let's explain your answer.

③ How do you think in order to find the number of standing women? Let's explain your answer.

Reflect Connect

Problem

In the target game, what is the tendency of getting on the target in the 1st and 2nd turn?

Name	Tatsuya	Mamoru	Aiko	Yuki	Takeshi	Kana	Atsushi	Kokoro
1st turn	◎	○	◎	◎	○	△	○	◎
2nd turn	◎	◎	◎	◎	△	○	○	○
Name	Yuki	Shizuka	Kazue	Rika	Yuko	Maya	Ken	Yui
1st turn	◎	◎	△	△	○	◎	◎	○
2nd turn	○	◎	△	△	○	◎	◎	◎
Name	Keiko	Manabu	Airi	Wakana	Daisuke	Nami	Yoshiharu	Takahiro
1st turn	○	○	◎	△	○	○	◎	△
2nd turn	○	◎	◎	○	◎	△	◎	○
Name	Yuma	Sota	Kazuhiro	Hiroki	Noriko	Toshihiro		
1st turn	◎	○	△	◎	○	◎		
2nd turn	○	○	○	○	◎	◎		

It is hard to see the numbers of ◎, ○, and △ on each turn.

Nanami

As for the 1st table, I understand the person and the tendency of getting on the target but it is not easy to see the numbers of ◎, ○, and △.

If another table is made, will it be easy to see? Let's make that.

Yui

[Target]

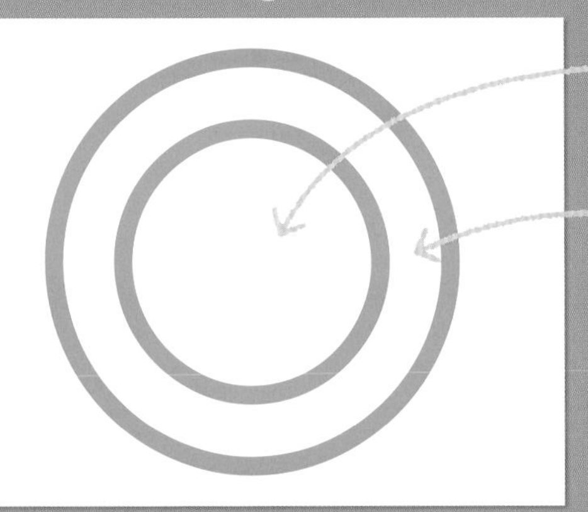

If you hit here ◎.

If you hit here ○.

If you do not hit △.

		2nd turn			Total
		◎	○	△	
1st turn	◎	9	4	0	13
	○	5	4	2	11
	△	1	3	2	6
Total		15	11	4	30

It is easy to understand because the 1st and 2nd turn are placed vertically and horizontally respectively.

Summary:

- The largest number of children hit ◎ on both 1st and 2nd turn.
- The second largest number of children hit ○ on 1st turn and ◎ on 2nd turn.
- There are no children with ◎ on 1st turn and △ on 2nd turn.

Daiki

Can I make such a table as I made before?

		Goldfish		Total
		Having	Not having	
Small birds	Having			
	Not having			
Total				

Want to connect

As the number of items increases, can I make a table that is easy to see by adding columns and rows?

Hiroto

Supplementary Problems

1 Large Numbers

p.10~p.22

1 Let's write the numerals for the following numbers.

① 8 billion 240 million 700 thousand.
② 56 billion 930 thousand.
③ 7 trillion 902 billion 140 million 300 thousand.
④ 48 trillion 8 billion 400 million.

2 Let's write the numerals for the following numbers.

① The number that gathers 305 sets of 100 million.
② The sum of 47 sets of 1 billion and 6900 sets of 10 thousand.
③ The number that gathers 321 sets of 1 trillion.
④ The sum of 5 sets of 10 trillion and 980 sets of 100 million.

3 Let's fill in the □ with a number.

① 38 billion gathers □ sets of 1 billion.
② The number that is 1 million smaller than 100 million is □.
③ 29 trillion gathers □ sets of 1 trillion.
④ 1 trillion gathers □ sets of 10 billion.

4 Let's write the following numbers.

① 10 times 30 thousand.
② 100 times 5 billion.
③ $\frac{1}{10}$ of 60 million.
④ 100 times 200 million.
⑤ $\frac{1}{10}$ of 40 billion.
⑥ 10 times 7 billion.
⑦ 10 times 80 million.
⑧ $\frac{1}{10}$ of 5 trillion.

5 Let's write appropriate numbers for Ⓐ ~ Ⓓ on the following number lines.

①
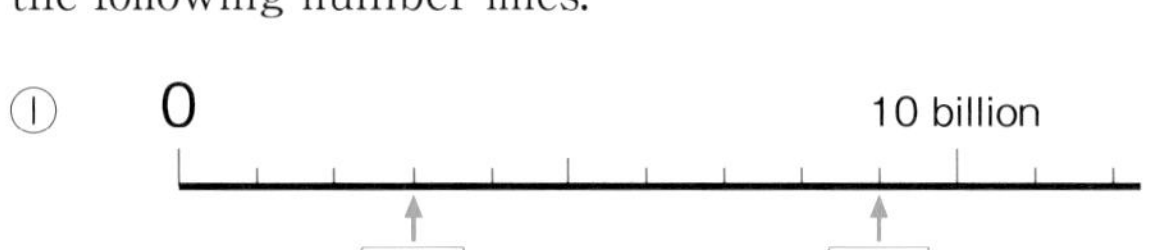

②
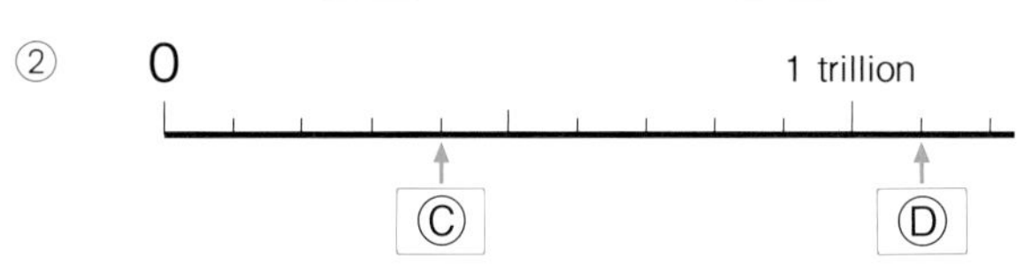

6 Let's fill in the □ with appropriate inequality signs.

① 5163800 □ 47291000
② 7243000 □ 7432000
③ 38725000 □ 38270000
④ 164309000 □ 164306000

7 Let's make various whole numbers using the five numerals: 0, 1, 2, 3, 3.

① Let's make the largest number.

② Let's make the smallest number.

8 Let's find the following sums.

① The sum of 3 billion 500 million and 5 billion 700 million.

② The sum of 2 trillion and 6 trillion.

③ 3 million 850 thousand + 2 million 560 thousand.

④ 17 billion 900 million + 24 billion 100 million.

9 Let's find the following differences.

① The difference between 4 billion 300 million and 2 billion 500 million.

② The difference between 16 trillion and 9 trillion.

③ 7 million 160 thousand − 3 million 380 thousand.

④ 26 billion 400 million − 19 billion 700 million.

10 Let's find the following products.

① 2 million 350 thousand × 4

② 6 billion 700 million × 3

③ 45 trillion × 10

11 Let's find the following quotients.

① 4 million 320 thousand ÷ 2

② 68 billion ÷ 10

③ 50 trillion ÷ 5

2 Line Graph

p.23~p.35

1 Look at the line graph below and let's answer the following questions about how the temperature changes in one day.

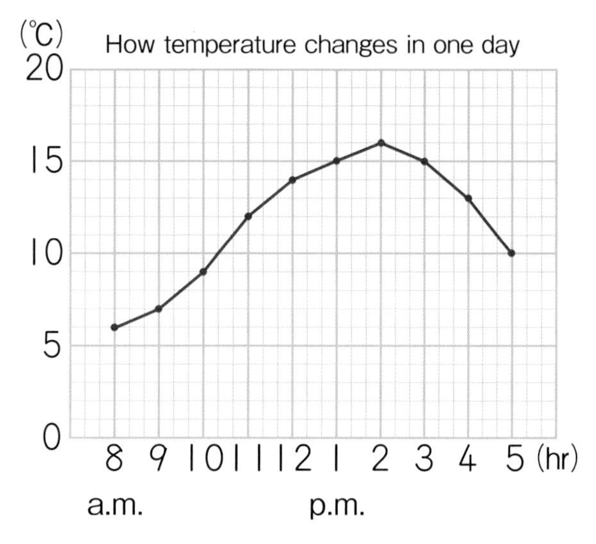

① What was the temperature in ℃ at 10 a.m.?

② What time was the temperature 15 ℃ ?

③ What was the highest temperature in ℃ and what time?

④ Between what consecutive time did the temperature rise the most?

⑤ By how many ℃ did the temperature fall between 4 p.m. and 5 p.m.?

⑥ What is the difference in temperature in ℃ between 9 a.m. and 1 p.m.?

2 Considering situations Ⓐ~Ⓓ, which is good to be represented by a line graph?

Ⓐ Favorite fruits of my classmates.

Ⓑ Body temperature taken every two hours when you have a cold.

Ⓒ The temperature of 6 places in school at 10 a.m.

Ⓓ Body weight per month.

3 The following table shows how the temperature changes in one year in Shizuoka City, Shizuoka Prefecture.

Let's draw a line graph.

How temperarture changes in Shizuoka City in one year

Month	1	2	3	4	5	6	7	8	9	10	11	12
Temperature (℃)	7	8	9	15	20	22	28	28	24	19	14	8

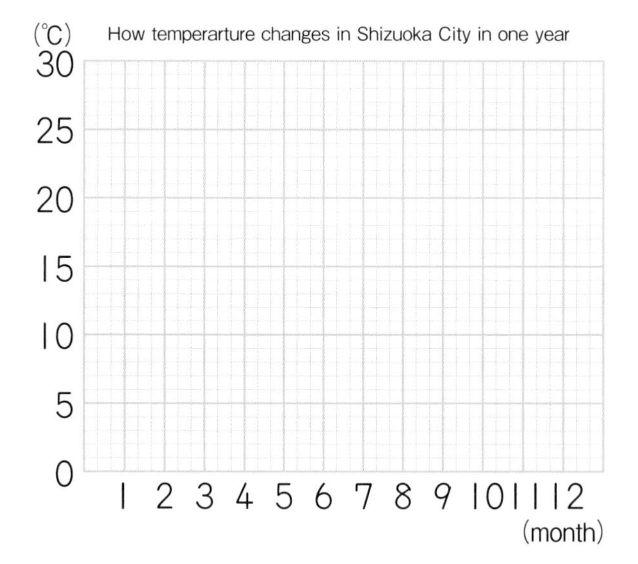

4 Division by 1-digit Number

p.42~p.57

1 Let's solve the following calculations.

① 40 ÷ 2 ② 60 ÷ 2
③ 800 ÷ 4 ④ 900 ÷ 3

2 Let's solve the following calculations in vertical form.

① 62 ÷ 7 ② 32 ÷ 5
③ 57 ÷ 8 ④ 30 ÷ 6
⑤ 65 ÷ 9 ⑥ 28 ÷ 4
⑦ 8 ÷ 6 ⑧ 9 ÷ 3

3 We want to divide 55 sheets of colored paper such that 7 children receive the same number. How many sheets of paper will each child receive and how many sheets of paper will remain?

4 Let's solve the following calculations in vertical form.

① 32 ÷ 2 ② 96 ÷ 4
③ 78 ÷ 6 ④ 98 ÷ 7
⑤ 75 ÷ 5 ⑥ 60 ÷ 4
⑦ 76 ÷ 2 ⑧ 84 ÷ 6

5 Let's solve the following calculations in vertical form. Also, let's confirm the answer.

① 75 ÷ 4 ② 53 ÷ 3
③ 86 ÷ 3 ④ 74 ÷ 6

6 Let's solve the following calculations in vertical form.

① 85 ÷ 4　② 66 ÷ 5
③ 94 ÷ 7　④ 73 ÷ 6
⑤ 61 ÷ 3　⑥ 83 ÷ 4

7 We want to divide 63 candies such that each child receives 5 candies. How many children can receive the candies? How many candies will remain?

8 Let's solve the following calculations in vertical form.

① 314 ÷ 2　② 447 ÷ 3
③ 775 ÷ 5　④ 588 ÷ 4
⑤ 825 ÷ 3　⑥ 912 ÷ 8
⑦ 891 ÷ 3　⑧ 984 ÷ 4

9 We want to divide 417 sheets of colored paper such that 3 groups receive the same number. How many sheets of paper will each group receive?

10 Let's solve the following calculations in vertical form.

① 196 ÷ 7　② 783 ÷ 9
③ 199 ÷ 4　④ 116 ÷ 2
⑤ 779 ÷ 8　⑥ 145 ÷ 3
⑦ 289 ÷ 5　⑧ 194 ÷ 3

11 Karen is reading a book with 224 pages. She reads 8 pages a day. How many days will it take her to finish?

12 Let's solve the following calculations in vertical form.

① 840 ÷ 6　② 810 ÷ 3
③ 915 ÷ 7　④ 654 ÷ 5
⑤ 618 ÷ 2　⑥ 836 ÷ 4
⑦ 736 ÷ 7　⑧ 975 ÷ 9

13 In the division on the right, a quotient is written from the tens place. Let's fill in the □ with all appropriate numbers from 1 to 9.

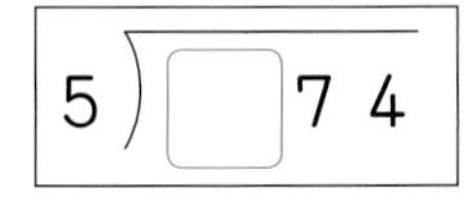

5 Angles

p.58~p.73

1 Let's fill in the □ with numbers.

① 1 right angle = □°
② □ right angles = 180°
③ 4 right angles = □°

2 How many degrees are the following angles?

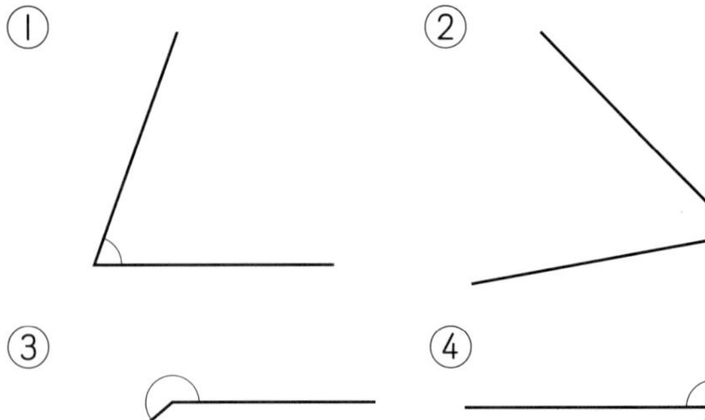

3 Two straight lines intersect as shown below.

How many degrees are the angles Ⓐ, Ⓑ, and Ⓒ?

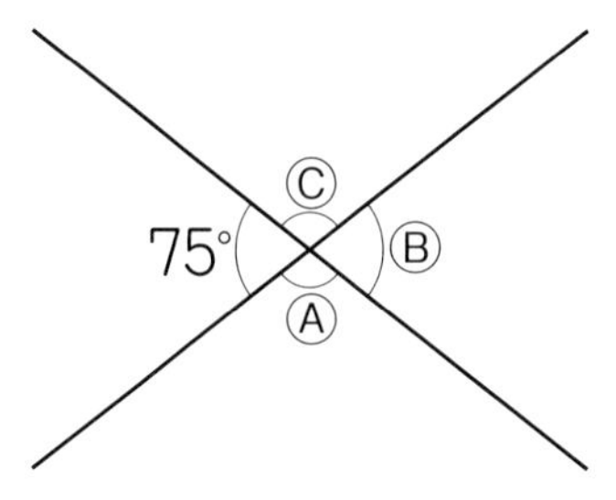

4 Let's draw angles with the following degrees.

① 65°

② 200°

③ 320°

5 As shown below, two triangle rulers are combined to make angles. How many degrees are angles Ⓐ, Ⓑ, and Ⓒ?

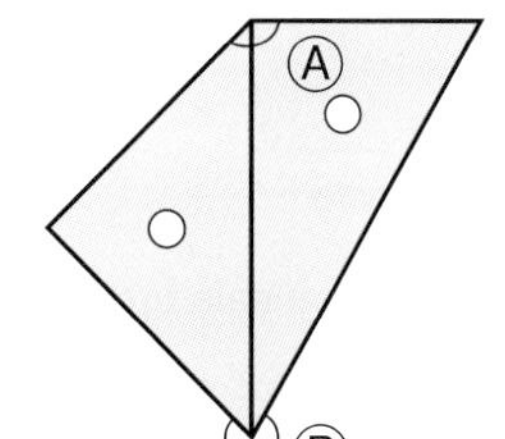

②

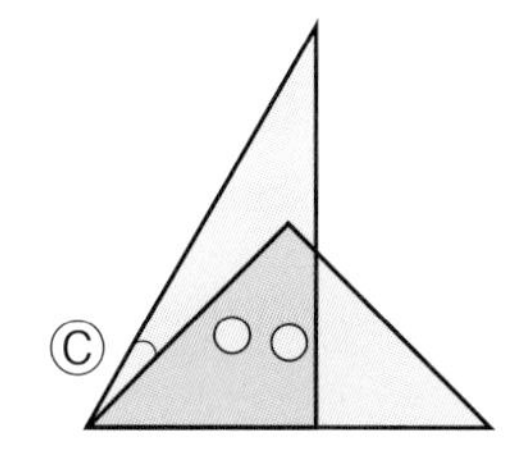

6 Perpendicular, Parallel and Quadrilaterals

p.74~p.101

1 In the following diagram, which are perpendicular straight lines?

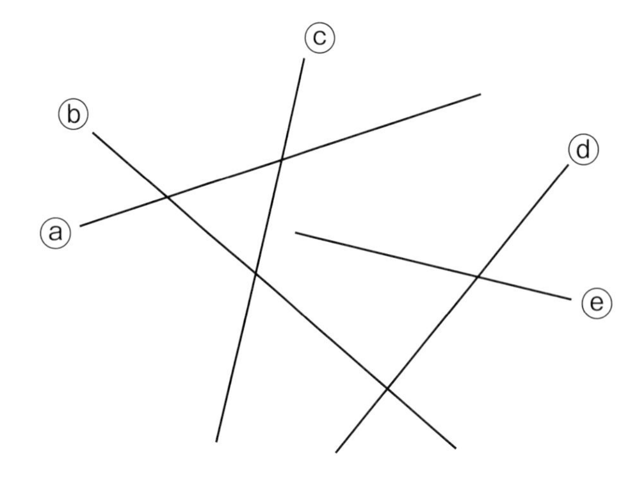

2 Through point A, let's draw a perpendicular straight line to straight line ⓐ.

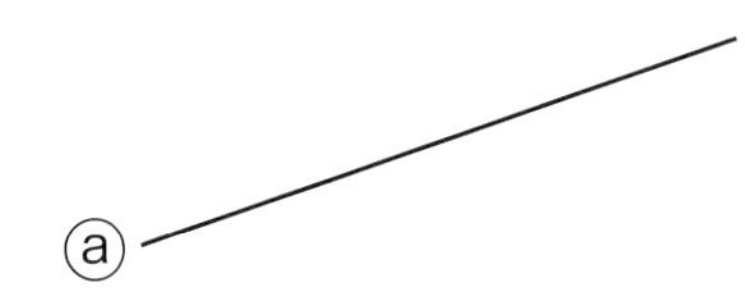

3 In the following diagram, which are parallel straight lines?

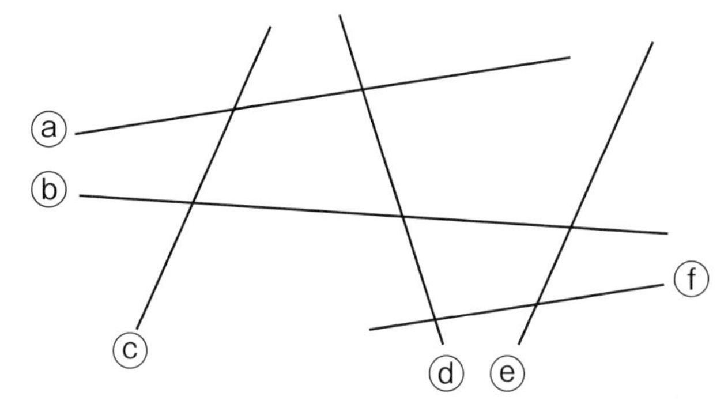

4 In the following figure, straight lines ⓐ and ⓑ are parallel.

How many degrees are angles Ⓐ and Ⓑ?

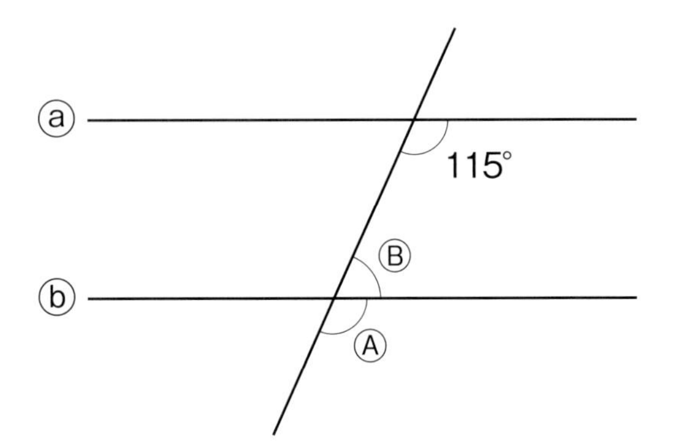

5 Through point A, let's draw a parallel straight line to straight line ⓐ.

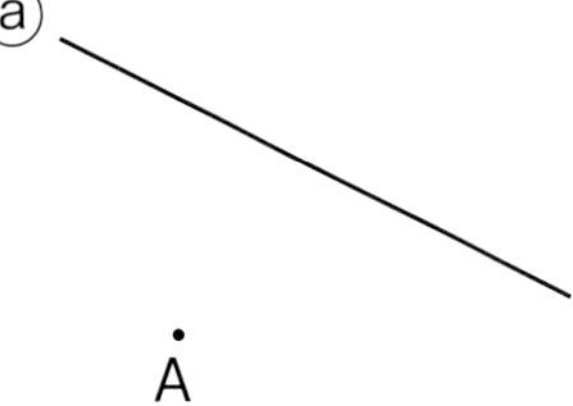

6 Let's answer the following questions about the parallelogram shown on the right.

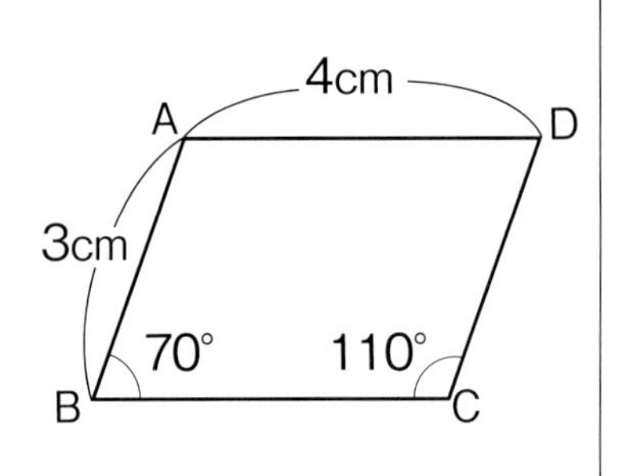

① How many cm is the length of side BC?

② How many degrees is the size of angle D?

7 Let's draw the following parallelogram.

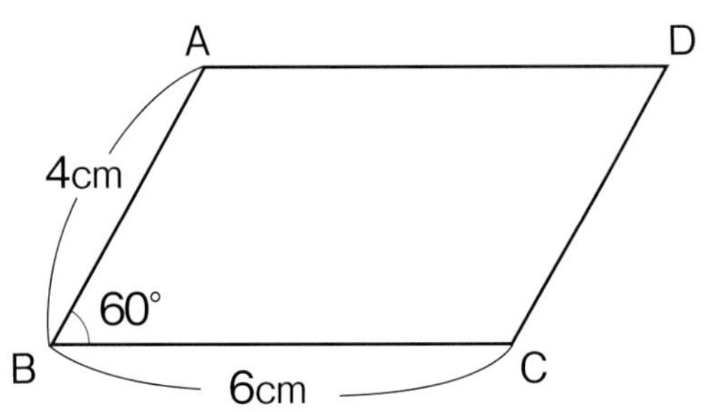

8 Let's answer the questions about the following rhombus.

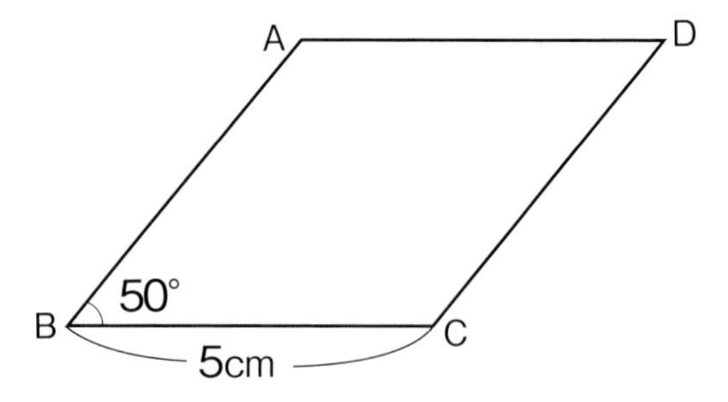

① Which side is parallel to side AB?

② How many degrees is the size of angle D?

③ Let's draw the same rhombus as this one.

9 Let's choose all quadrilaterals that have the properties ①~④ shown below.

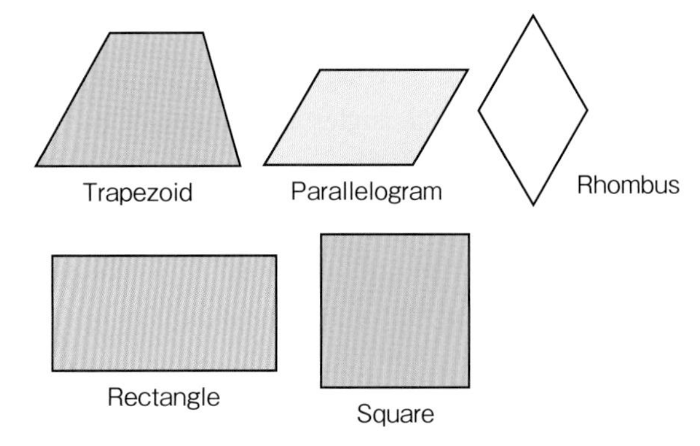

① A quadrilateral where two diagonals intersect perpendicularly.

② A quadrilateral where two diagonals have the same length.

③ A quadrilateral where two diagonals intersect perpendicularly and have the same length.

④ A quadrilateral in which two diagonals are divided into 2 equal parts at the point where they intersect.

10 In the following diagram, what kind of quadrilateral is drawn by connecting the points A → B → C → D → A with straight lines?

① 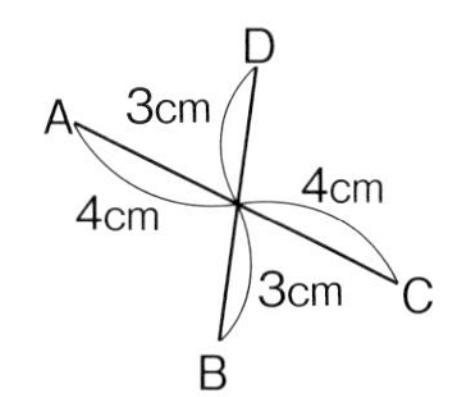

② 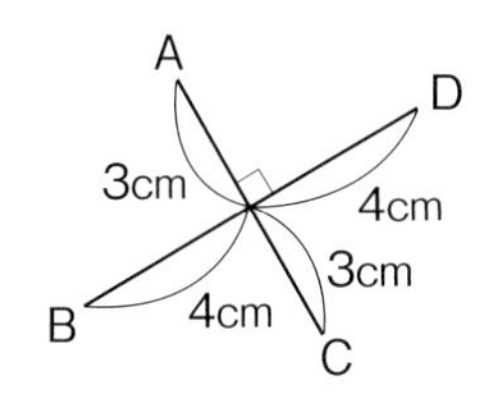

7 Division by 2-digit Number

p.102~p.119

1 Let's solve the following calculations.

① 80 ÷ 40 ② 90 ÷ 30

③ 160 ÷ 20 ④ 70 ÷ 30

⑤ 430 ÷ 50 ⑥ 290 ÷ 70

2 We want to divide 230 balls such that 1 box contains 40 balls. How many boxes will be filled? How many balls will remain?

3 Let's solve the following calculations in vertical form.

① 84 ÷ 42 ② 48 ÷ 12

③ 66 ÷ 22 ④ 29 ÷ 13

⑤ 67 ÷ 21 ⑥ 96 ÷ 31

4 Let's solve the following calculations in vertical form.

① 52 ÷ 13 ② 84 ÷ 28

③ 90 ÷ 37 ④ 75 ÷ 15

⑤ 72 ÷ 14 ⑥ 85 ÷ 17

5 We want to cut a ribbon that is 65 cm long into pieces with a length of 16 cm. How many pieces with a length of 16 cm can we get? How many cm will remain?

6 Let's solve the following calculations in vertical form.

① 210 ÷ 42 ② 243 ÷ 81

③ 218 ÷ 38 ④ 343 ÷ 53

⑤ 539 ÷ 74 ⑥ 282 ÷ 47

⑦ 309 ÷ 31 ⑧ 742 ÷ 77

7 We want to divide 135 sheets of colored paper such that 18 children receive the same number of sheets. How many sheets of paper will each child receive? How many sheets will remain?

8 Let's solve the following calculations in vertical form.

① 450 ÷ 18 ② 551 ÷ 46

③ 855 ÷ 57 ④ 391 ÷ 23

⑤ 432 ÷ 27 ⑥ 965 ÷ 36

9 Let's solve the following calculations in vertical form.

① 924 ÷ 46 ② 648 ÷ 16

③ 963 ÷ 48 ④ 729 ÷ 24

⑤ 765 ÷ 76 ⑥ 688 ÷ 34

10 Let's fill in each □ with a number.

① 3000 ÷ 500 = □ ÷ 5
= □

② 2800 ÷ 700 = 28 ÷ □
= □

③ 125 ÷ 25 = □ ÷ 50
= □

8 Round Numbers

p.124~p.138

1 Let's round off 5□83 to the nearest thousand round number.

① Let's write all the numerals that apply in □ when the number becomes about 5000.

② Let's write all the numerals that apply in □ when the number becomes about 6000.

2 Let's round off the following numbers to the nearest thousand round number. Also, to the nearest ten thousand round number.

① 75162 ② 90735

③ 428460 ④ 833649

⑤ 360549 ⑥ 299130

3 Let's round off the following numbers from the highest first digit round number. Also, from the highest two digits round number.

① 845 ② 1594 ③ 6064

④ 43702 ⑤ 61842 ⑥ 281540

4 Let's round down the following numbers from the highest two digits round number. Also, round up from the highest first digit round number.

① 54632 ② 8193

③ 33370 ④ 9756

5 The following numbers represent the round number for rounding off whole numbers to the nearest place written in the [ ]. What is the range for those whole numbers by using the term "greater than or equal to" and "less than"?

① 20 [to the nearest ten]

② 280 [to the nearest ten]

③ 600 [to the nearest hundred]

④ 3000 [to the nearest thousand]

9 Arrangement of Data

p.139~p.145

1 The following table summarizes the injuries that occurred at school last month, focusing on the places where the injuries occurred and the kind of injuries. Let's answer the questions about this table.

Place and kind of injury (children)

Kind / Place	Cut	Bruise	Scratch	Sprain	Total
Playground	1	4		1	
Corridor	2	3	2	0	
Classroom	4	1	3	0	
Gymnasium	1	2	4	2	
Stairs	0	1	2	0	
Total	8		16		

① What is the number of children for the following?

ⓐ The number of children with bruises at the corridor.

ⓑ The number of children with scratches at the gymnasium?

② How many children had a scratch at the playground?

③ Let's fill in the blanks wih numbers in the table.

④ What kind of injury occurred most frequently?

⑤ Where did injuries occur most frequently?

⑥ How many children were injured in total at school last month?

2 Haruma investigated whether his classmates had a dog or a cat. The following information was gathered:

- 24 children have a dog.
- 11 children have a dog and a cat.
- 8 children have a cat but no dog.
- 18 children do not have a cat.
- Haruma's class has 37 children.

Let's answer the following problems.

① Let's write down in the table the numbers of children you know. Also, let's fill in the blanks with numbers of children.

Animals that children have (children)

		dog		Total
		○	×	
cat	○			
	×			
Total				

(○…have ×…do not have)

② How many children do not have either a dog or a cat?

③ How many children only have a dog?

④ How many children do not have a dog?

⑤ How many children have a cat?

1 Large Numbers

1 ① 8240700000 ② 56000930000
③ 7902140300000 ④ 48008400000000

2 ① 30500000000 ② 47069000000
③ 321000000000000 ④ 50098000000000

3 ① 38 ② 99 million ③ 29 ④ 100

4 ① 300 thousand ② 500 billion ③ 6 million
④ 20 billion ⑤ 4 billion ⑥ 70 billion
⑦ 800 million ⑧ 500 billion

5 ① Ⓐ 3 billion Ⓑ 9 billion ② Ⓒ 400 billion
Ⓓ 1 trillion 100 billion

6 ① < ② < ③ > ④ >

7 ① 33210 ② 10233

8 ① 9 billion 200 million ② 8 trillio
③ 6 million 410 thousand ④ 42 billion

9 ① 1 billion 800 million ② 7 trillion
③ 3 million 780 thousand ④ 6 billion 700 million

10 ① 9 million 400 thousand ② 20 billion 100 million
③ 450 trillion

11 ① 2 million 160 thousand ② 6 billion 800 million
③ 10 trillion

2 Line Graph

1 ① 9℃ ② 1 p.m. and 3 p.m. ③ 16℃, 2 p.m.
④ Between 10 a.m. and 11 a.m. ⑤ 3℃ ⑥ 8℃

2 Ⓑ, Ⓓ

3

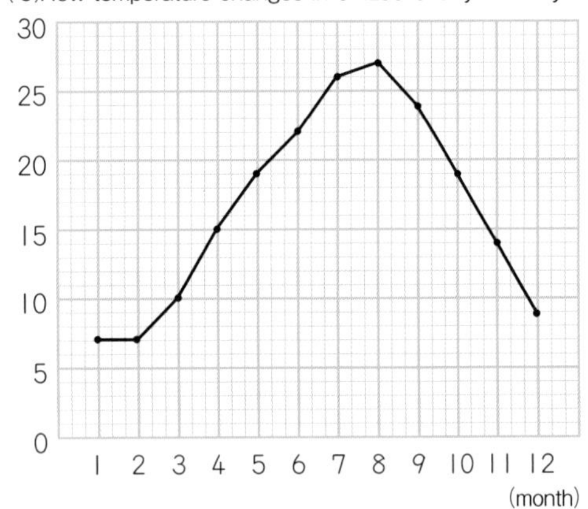

4 Division by 1-digit Number

1 ① 20 ② 30 ③ 200 ④ 300

2 ① 8 remainder 6 ② 6 remainder 2 ③ 7 remainder 1
④ 5 ⑤ 7 remainder 2 ⑥ 7 ⑦ 1 remainder 2 ⑧ 3

3 55 ÷ 7 = 7 remainder 6 Each child receives 7 sheets and 6 sheets remain.

4 ① 16 ② 24 ③ 13 ④ 14 ⑤ 15 ⑥ 15
⑦ 38 ⑧ 14

5 ① 18 remainder 3, 4 × 18 + 3 = 75
② 17 remainder 2, 3 × 17 + 2 = 53
③ 28 remainder 2, 3 × 28 + 2 = 86
④ 12 remainder 2, 6 × 12 + 2 = 74

6 ① 21 remainder 1 ② 13 remainder 1
③ 13 remainder 3 ④ 12 remainder 1
⑤ 20 remainder 1 ⑥ 20 remainder 3

7 63 ÷ 5 = 12 remainder 3 12 children, 3 candies remain.

8 ① 157 ② 149 ③ 155 ④ 147 ⑤ 275
⑥ 114 ⑦ 297 ⑧ 246

9 417 ÷ 3 = 139 139 sheets

10 ① 28 ② 87 ③ 49 remainder 3 ④ 58
⑤ 97 remainder 3 ⑥ 48 remainder 1
⑦ 57 remainder 4 ⑧ 64 remainder 2

11 224 ÷ 8 = 28 28 days

12 ① 140 ② 270 ③ 130 remainder 5
④ 130 remainder 4 ⑤ 309 ⑥ 209
⑦ 105 remainder 1 ⑧ 108 remainder 3

13 1, 2, 3, 4

5 Angles

1 ① 90 ② 2 ③ 360

2 ① 70° ② 55° ③ 220° ④ 300°

3 Ⓐ 105° Ⓑ 75° Ⓒ 105°

4

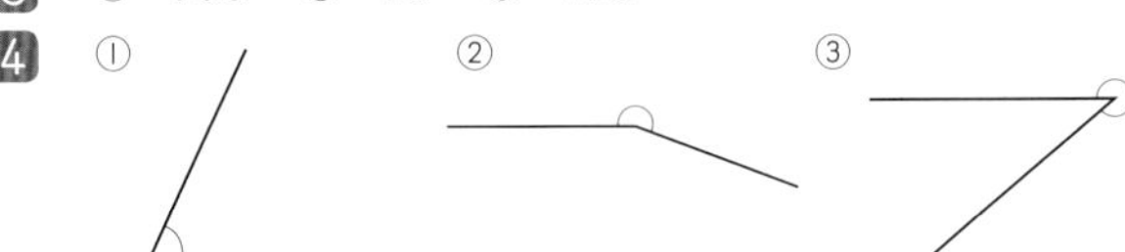

5 ① Ⓐ 135° Ⓑ 285° ② Ⓒ 15°

6 Perpendicular, Parallel and Quadrilaterals

1 straight line ⓐ and straight line ⓓ, straight line ⓒ and straight line ⓔ

2

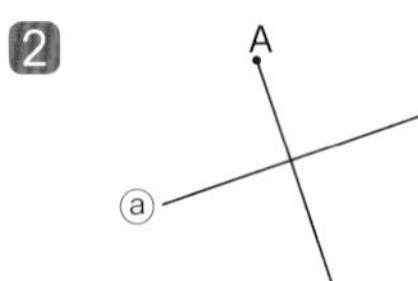

3 straight line ⓐ and straight line ⓕ, straight line ⓒ and straight line ⓔ

4 Ⓐ 115° Ⓑ 65°

5

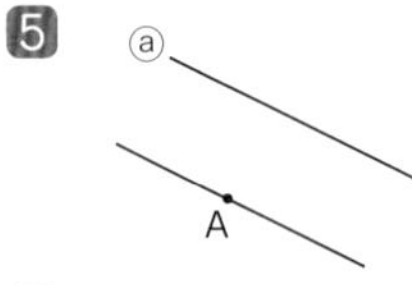

6 ① 4cm ② 70°

7 (Omitted)

8 ① sideDC ② 50° ③ (Omitted)

9 ① Rhombus, Square ② Rectangle, Square ③ Square
④ Parallelogram, Rhombus, Rectangle, Square

10 ① Parallelogram ② Rhombus

7 Division by 2-digit Number

1 ① 2 ② 3 ③ 8 ④ 2 remainder 10
⑤ 8 remainder 30 ⑥ 4 remainder 10

2 230 ÷ 40 = 5 remainder 30 5 boxes, 30 balls remain.

3 ① 2 ② 4 ③ 3 ④ 2 remainder 3 ⑤ 3 remainder 4
⑥ 3 remainder 3

4 ① 4 ② 3 ③ 2 remainder 16 ④ 5
⑤ 5 remainder 2 ⑥ 5

5 65 ÷ 16 = 4 remainder 1 4 pieces, 1 cm remains.

6 ① 5 ② 3 ③ 5 remainder 28 ④ 6 remainder 25
⑤ 7 remainder 21 ⑥ 6 ⑦ 9 remainder 30
⑧ 9 remainder 49

7 135 ÷ 18 = 7 remainder 9 7 sheets, 9 sheets remain.

8 ① 25 ② 11 remainder 45 ③ 15 ④ 17 ⑤ 16
⑥ 26 remainder 29

9 ① 20 remainder 4 ② 40 remainder 8
③ 20 remainder 3 ④ 30 remainder 9
⑤ 10 remainder 5 ⑥ 20 remainder 8

10 ① 30, 6 ② 7, 4 ③ 250, 5

8 Round Numbers

1 ① 0, 1, 2, 3, 4 ② 5, 6, 7, 8, 9

2

	①	②	③
To the nearest thousand	75000	91000	428000
To the nearest ten thousand	80000	90000	430000

	④	⑤	⑥
To the nearest thousand	834000	361000	299000
To the nearest ten thousand	830000	360000	300000

3

	①	②	③
First digit	800	2000	6000
Two digits	850	1600	6100

	④	⑤	⑥
First digit	40000	60000	300000
Two digits	44000	62000	280000

4

	①	②	③	④
Round down	54000	8100	33000	9700
Round up	60000	9000	40000	10000

5 ① Greater than or equal to 15 and less than 25
② Greater than or equal to 275 and less than 285
③ Greater than or equal to 550 and less than 650
④ Greater than or equal to 2500 and less than 3500

9 Arrangement of Data

1 ① ⓐ 3 children ⓑ 4 children ② 5 children
③ Place and kind of injury (children)

Place \ Kind	Cut	Bruise	Scratch	Sprain	Total
Playground	1	4	5	1	11
Corridor	2	3	2	0	7
Classroom	4	1	3	0	8
Gymnasium	1	2	4	2	9
Stairs	0	1	2	0	3
Total	8	11	16	3	38

④ Scratch ⑤ Playground ⑥ 38 children

2 ① Animals that children have (children)

		Dog		Total
		○	×	
Cat	○	11	8	19
	×	13	5	18
Total		24	13	37

② 5 children ③ 13 children ④ 13 children
⑤ 19 children

Words and symbols from this book

Various quadrilaterals ▼ will be used in pages 75 and 85

Angles ▼ will be used in pages 60 and 61

Memo

Memo

Memo

Editorial for English Edition:

Study with Your Friends, Mathematics for Elementary School

4th Grade, Vol.1, Gakko Tosho Co.,Ltd., Tokyo, Japan [2020]
